ENERGY SPACE TIME

Introducing Energy Metrics of Interactions in Five Dimensions

ENERGY SPACE TIME

Introducing Energy Metrics of Interactions in Five Dimensions

Fabio Cardone
National Research Council of Italy, Italy

Domenico Bassani
Sidom LLC, Italy

World Scientific

NEW JERSEY • LONDON • SINGAPORE • BEIJING • SHANGHAI • HONG KONG • TAIPEI • CHENNAI • TOKYO

Published by

World Scientific Publishing Co. Pte. Ltd.
5 Toh Tuck Link, Singapore 596224
USA office: 27 Warren Street, Suite 401-402, Hackensack, NJ 07601
UK office: 57 Shelton Street, Covent Garden, London WC2H 9HE

Library of Congress Control Number: 2025037065

British Library Cataloguing-in-Publication Data
A catalogue record for this book is available from the British Library.

On the cover "Asymmetric Energy", oil painting on canvas by Mario Cardone, 2005.

ENERGY SPACE TIME
Introducing Energy Metrics of Interactions in Five Dimensions

ISBN 978-981-98-2162-4 (hardcover)
ISBN 978-981-98-2163-1 (ebook for institutions)
ISBN 978-981-98-2164-8 (ebook for individuals)

For any available supplementary material, please visit
https://www.worldscientific.com/worldscibooks/10.1142/14537#t=suppl

Desk Editors: Nambirajan Karuppiah/Rok Ting Tan

Typeset by Stallion Press
Email: enquiries@stallionpress.com

To the Memory of the dear colleagues deceased in the Science

Mauro Francaviglia († 2013)
who contributed to explore the mathematics of the energy metrics for the interactions

Roberto Mignani († 2018)
who saw the need for the physical extended fifth dimension in the energy metrics for interactions

Eliano Pessa († 2020)
who proposed an appropriate dimensional homogeneity space constant for the fifth dimension

Giovanni Cherubini († 2022)
who measured the neutrons emitted by Deformed Space Time reactions

Ludwik Kostro († 2024)
who calculated the constant named after him which gives the upper limit for the recursivity of the energy metrics for interactions

Sergio Benenti († 2025)
who contributed to explore the mathematics of the Ricci Flow of the energy metrics for the interactions

Valter Sala († 2025)
who was the first to build the ultrasound machines producing nuclear metamorphosis at macroscopic level

Foreword

The experiment is an operation aimed at knowing
an unknown truth, but it is a risk too.

Nigel Kneale, Penguin Books, 1959,
The Quatermass Experiment

The geometry of interactions is regarded as the space time geometry of the energy exchange between interacting bodies.

The triple role of the energy is: phenomenological variable of the geometric representation, fifth coordinate of geometric representation together with the three coordinates of the space and the coordinate of the time, finally as a quantity subject to a conservation of its total value in the physical process under consideration, which is the interaction.

This is the first great cornerstone of our work.

The second cornerstone is the evidence that there are two types of violation of the local Lorentz invariance, which works as a test of the flat or non flat nature of the space-time.

- The first violation is of kinematic type and leads from Galilean transformations to Lorentz transformations.
- The second violation is of the geometric type and leads in four dimensions from Lorentz transformations to deformed Lorentz transformations. And this is a fact for us.
- Third and last cornerstone, the geometric Lorentz violation has a general as well fundamental character, since it involves the existence of a privileged direction.

The concept of privileged direction completely replaces the concept of privileged reference system. So the privileged reference system does not exist, but the privileged direction is a physical reality, which can be verified both at the terrestrial level and at the cosmic level.

Here are, all the necessary assembly pieces for our work. Now, we can say what is the extreme frontier of this work: to link the energy with the geometric violation of the local Lorentz invariance.

Einstein changed his mind introducing within the conservation of the total energy also the energy equivalent to the inertia, or rather the mass even if he used the transverse mass, and in this way he solved the main problem of the Newtonian mechanics, the violation of the total energy conservation in the inverse inelastic collision, but effectively this was not a direct solution rather a change of the point of view. Similarly, we have introduced the deformation of the space time as dependent on the energy of the interaction as a tool to guarantee the conservation of total energy too.

So, we too have changed our point of view, but only that our change is of a geometrical type, instead the Einstein's one is of a mechanical type. Effectively, also we have circumvented the problem by changing our point of view, like did Einstein. Since in our case in all nuclear metamorphosis reactions there is a clear violation of the total energy conservation: where does this missing energy end up, fact evidenced in the absence of gamma radiation from the nucleus? We argue that the missing energy, which is a difference in the measured energy, is absorbed by the deformation of space and time which from flat becomes deformed.

We privileged a book written in the Bohr way or better like Infeld and Einstein, using the physical concepts more than mathematical formalism ones and we do not fear entering heavily towards the summit, the pinnacle, of the foundations of the 'maxima' systems: starting from energy conservation and arriving to Copernican principle.

Warning

We warn the readers about the issue of technology. In some parts of this book describing experiments the reader could find the text containing omissions about both technological devices and technological protocols. Our choice was to describe the science of fundamental interactions and their geometrical descriptions as natural laws. But natural laws are not eligible and suitable to be patented, as well known after the Patent Convention Treaty, Paris 1976 and subsequent upgrades.

When speaking of science everybody understanding it, can perform his own technology which can be subjected to intellectual property.

Preface

Loquere, Domine, quia audit servus tuus
Bible Old Testament, Daniel Book 3, 9

Incerta et occulta sapientiae tuae manifestasti mihi
Liber Psalmorum 50, 8, Psalm 102

Nunc dimittis servum tuum, Domine,
secundum verbum tuum in pacem:
quia viderunt oculi mei salutare tuum
Luca 2, 29–30

Without physical meaning, mathematics in physics is a significant void. But how can the signifier of mathematical language be made to embody the logical properties of the meaning that is physics?

The difficulty is that the logic of physical nature is not human logic, whereas mathematics comes from human logic. So, the logic of the signifier is originally alien to the logic of the meaning: this is the true profound problem of theoretical physics. We wonder how we can represent in mathematical language, which is illogical to nature, the physical laws that instead follow the logic of nature. The solution is to adapt the signifier to the meaning and its logic. To do this, it is first necessary to make some warning considerations.

We learn the language of mathematics independently of the language of physics and we bring ourselves into thinking that the mathematical representation of the physical law carries the same logical power as mathematics itself: a mistake!

Mathematics only gives the illusion of offering to the physical law the deductive power of mathematical language. Instead, the connection between physical laws is determined only by the logic of nature, and to enter this logic one must first understand it and then comprehend it. Only then can one look for a mathematical signifier and, if it is not found in known mathematics, adapt the available mathematical language to the properties of the laws of nature from which the physical laws descend. Let us recall here that the laws of nature are the interactions, from which come the physical laws which describe the phenomenology that is measured in experiments.

Now in order to proceed towards the solution of the problem posed of adapting the mathematical signifier to the physical meaning and its logic, let us examine the two fundamental tools of reasoning: Deduction and Induction. Deduction and Induction must necessarily interpenetrate each other in both directions of the process of any logical physical reasoning and we now give their exposition.

Induction starts with the experiment, which leads to the measurement, which gives us the phenomenology, which gives us the physical law from which we obtain the law of nature: the interaction.

This allows us to understand the logic of the nature of the interactions in an extensive sense.

Deduction starts with the theory, which leads to the first principles of interaction, that is, it gives us the law of nature, which leads to the physical law, which allows us to conceive the experiment, which gives us the measurement, which leads to the phenomenology of the interaction.

This allows us to understand the logic of the nature of the interactions in an intensive sense.

Let us briefly note that phenomenology includes the repetitiveness of a measurement and with it its statistics.

The balanced use of Induction and Deduction has led us in previous works to conclude that the essential logical characteristic of the laws of nature is in their anisotropy and asymmetry. We therefore concluded that the mathematical language for describing the laws of nature, i.e. interactions, could not be restricted to the use of simple flat or non-flat geometries of a single type, for example either Minkowskian or Riemannian. We looked for geometries able to describe any flat or non-flat situation for each interaction.

In conclusion, we have recognized that the logic of nature is not restrictive and requires a mathematical language highly adaptable to the needs of interactions.

Our operational choice was a complete geometrization of the interactions allowing us to implement their characteristics independently of the geometry itself.

Prof. Fabio Cardone
Ph.D. Eng. Domenico Bassani
Italy
June 2, 2025

Acknowledgements

The authors gratefully acknowledge the help and contributions by Ph.D. Giovanna Bassani for her help in reviewing the text and by Ph.D. Silvia Dari for her help in handling the figures.

The authors gratefully acknowledge the help and contributions by dear colleagues who examined the provisional text of the book:

Umberto Bartocci, Stefano Bellucci, Fabio Pistella and Giovanni Rastelli.

The authors gratefully acknowledge the help and contributions also by dear friends who read the preliminary draft of the book:

Davide Barale, Eugenio Brusa, Augusto Galleano and Daniele Schito.

The authors give their thanks to the Senior Editor Mrs. Tan Rok Ting, who took care of this book, but she also took care in the past of the book entitled 'Energy and Geometry', the forerunner of this volume.

Last but not the least, or rather the best, our dear colleague, Gianni Albertini with whom the authors are deeply indebted for his work in deep of scientific revision of the whole book. Our gratitude will last for ever.

About the Authors

Fabio Cardone, born in Chieti, Italy, in 1960, former Professor of Modern Physics at the Pontifical Universities Gregorian and St. Thomas 'Angelicum' in Rome, was Professor of General Physics at the University of L'Aquila and the University La Sapienza of Rome.

He holds courses in the field of relativity in some American universities, Madison (WI), Syracuse (NY), Maryland College Park (MD), MIT, and Danzig University in Poland.

He was a winner in 1985 of the Italian National Prize for Physics 'G. Galilei' and former Fellow of the Italian National Institute of High Mathematics (INDAM). As a former member of the Academy of Sciences of New York (NYAS), he was also conferred with the Medal for Theoretical Physics at Cambridge International Biographical Institute (2000). He conducts research in the physics of fundamental interactions and the theory of relativity. Since 2003, he has been a permanent member of the National Research Council of Italy (CNR).

Domenico Bassani, born in Turin, Italy, in 1959, has degrees in Mechanical Engineering, PhD in Mechatronics and RG in Electronics and Robotics at the Polytechnic University of Turin.

Since 2011, he has been conducting research in physics of the Deformed Space Time (DST), participating in the discovery of the matter nuclear metamorphoses in 2015 and in 2019 to the torsional antenna behaviour according to the DST 'angle law'.

He studied the exploitation of the DST reactions by ultrasound machines.

From 2022, he directed the design and the making of the reactor at Metamorphosys (RaM), as the first reactor producing DST reactions on a macroscopic level and large scale.

Contents

Part I

Energy Metrics

Chapter 1

Phenomenological Energy Metrics

The first step towards a representation of fundamental interactions in a geometrical framework including energy together space and time is to look for metrics depending on energy obtained through experiments on an energy range as wide as possible [1]. For this reason we selected four experiments, one for each interaction, which we list here below by interaction:

Electromagnetic (e.m.) waveguide experiment, propagation of evanescent mode in crossing a waveguide with variable section;

Gravitational clock rate experiment, time measure at different altitudes with respect to the ground;

Leptonic the decay lifetime of the meson K^0_S experiment at different energies;

Hadronic the proton antiproton annihilation at different energies, the CERN 'ramping run' experiment.

The analysis of the above listed experiments gave rise to phenomenological energy metrics in four dimensions for the coordinates

$$(t, x, y, z) \quad \text{or rather} \ (x_0, x_1, x_2, x_3)$$

and four-dimensional metrics g whose elements are expressed by function b^2, say

$$g = (b_0^2, b_1^2, b_2^2, b_3^2).$$

The metric tensor g has always been assumed diagonal and the metric elements are function all depending by energy E, and therefore in the analysis of the experiments we have used the following expression for the metrics:

$$g(E) = \mathrm{diag}(b_0^2(E), b_1^2(E), b_2^2(E), b_3^2(E)).$$

1.1 Waveguide Experiment

In order to derive the explicit form of the e.m. metric describing such non-local effects, we analyse the Cologne wave-guide experiments on the propagation of e.m. waves in conducting wave-guides with variable Section [2–4].

Let us consider the speed of an evanescent mode in crossing the reduced section wave-guide with length L which is plotted, in Fig. 1.1, as a function of L.

Indeed, the wave frequency ν in the guide with greater section was lower than the cutoff frequency ν_c of the guide with smaller section.

On account of the mode involved (TE_{10}), it is possible to assume that space is isotropic inside the smaller waveguide. Moreover, since the propagation speed u depends only on the ratio (b/b_0), see Eq. (1.1), we can also put $b_0^2 = 1$, i.e. the deformed metric is isochronous with the flat Minkowskian metric. Therefore, the metric to be considered is

$$g = \mathrm{diag}(1, -b^2, -b^2, -b^2). \tag{1.1}$$

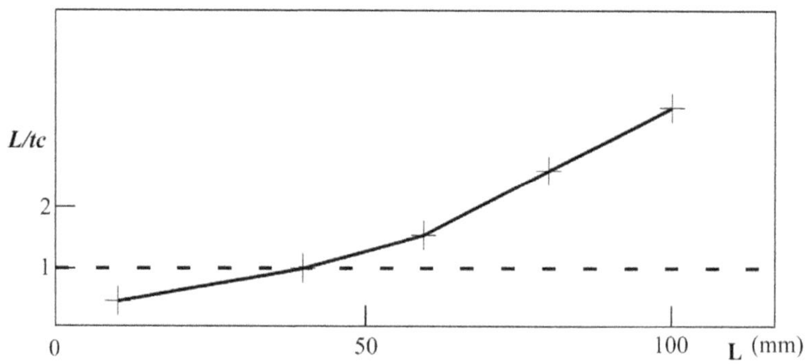

Fig. 1.1. Plot of the transmission velocity L/tc vs. the length L of the reduced waveguide of the Cologne experiment.

The expression of the propagation speed of the evanescent mode is assumed to be [1]

$$u = \frac{c}{b} \tag{1.2}$$

and this relation admits to deduce the values of the parameter b. The wave vector is given by the following Eq. (1.3), which includes the metric (1.1) and therefore has the form:

$$\tilde{k}_z^2 = \left(\frac{2\pi}{c}\right)^2 \nu^2 \left[\frac{1}{b^2} - \left(\frac{\nu_c}{\nu}\right)^2\right]. \tag{1.3}$$

In correspondence to the computed values of b, Eq. (1.3) permits to find the values of the wave vector component $\tilde{k}_z$ for the frequencies used in the experiments, i.e. $\nu = 8.70\,\text{GHz}$ and $\nu_c = 9.49\,\text{GHz}$.

The ratio ν_c/ν yields the condition whereby $\tilde{k}_z^2 \geq 0$ holds, i.e.

$$\frac{1}{b^2} > \left(\frac{\nu_c}{\nu}\right)^2 = 1.091. \tag{1.4}$$

Moreover, exploiting the classical theory of an evanescent mode, it is possible, from the knowledge of the length L of the reduced-section guide, to evaluate the output energies corresponding to given values of L and b. The exploited relation is

$$E = h\nu e^{-L/L_0}, \tag{1.5}$$

where h is the Planck constant, ν the photon frequency at the input of the reduced-section guide, and L_0 is the penetration length of the energy E.

Note that Eq. (1.5) does not mean at all that fractions of the photon energy are involved. As for the case of total internal reflection, the meaning of the above expression is simply that the energy lost by the signal after travelling a distance L inside the reduced waveguide

$$\Delta E = E_{\text{in}} - E = h\nu(1 - e^{-L/L_0}) \tag{1.6}$$

can be considered as the energy needed to deforming the space-time inside the barrier.

The length L_0 is just provided by the classical theory of the evanescent mode:

$$L_0 = \left(\frac{c}{2\pi}\right)\frac{1}{\sqrt{\nu_c^2 - \nu^2}}. \tag{1.7}$$

For the frequency values used in the experiments, L_0 is real because $\nu < \nu_c$, and it is $L_0 = 1.256\,\text{cm}$.

Figure 1.2 shows the behaviour of the parameter b^2 vs. E for values $u > c$. The interpolation curve has been obtained using the values of E and b^2 for which $\tilde{k}_z^2 \geq 0$. We give below the explicit functional form of the corresponding metric $g_{\text{e.m.}}(E)$, that is to be regarded as a mere phenomenological description of the e.m. interaction under deformed space-time conditions [5]:

$$g_{\text{e.m.}}(E) = \text{diag}(1, -b^2_{\text{e.m.}}(E), -b^2_{\text{e.m.}}(E), -b^2_{\text{e.m.}}(E))$$
$$b^2_{\text{e.m.}}(E) = \begin{cases} (E/E_{0,\text{e.m.}})^{1/3} & 0 \leq E < E_{0,\text{e.m}} = 4.5 \pm 0.2\mu eV \\ 1, & E_{0,\text{e.m}} \leq E \end{cases}. \tag{1.8}$$

The threshold energy $E_{0,\text{e.m.}}$ is the energy value at which the metric parameters are constant, i.e. the metric becomes flat, Minkowskian, and the electromagnetic interaction is fully derivable from a potential. Note that $g_{\text{e.m.}}(E)$ is a *sub-Minkowskian* metric, namely it

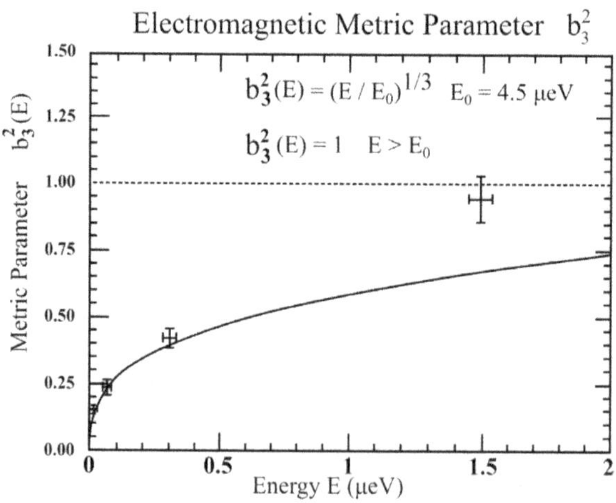

Fig. 1.2. Fit of the electromagnetic metric parameter $b_3^2 = b^2(\text{E})$ vs. the energy of the output signal.

reaches the flat space time, the Minkowskian limit, from below for energies $E < E_{0,\text{e.m.}}$.

1.2 Clock Rates Experiments

In order to describe the gravitational interaction in terms of an energy-dependent metric, the experimental input is provided by the measurements carried out by Alley [6] in 1975 and by Meystre and Scully in 1983 (analysed in [7]) on the relative rates of clocks in the gravitational field. In such experiments, the rate of cesium beam atomic clocks, raised to a higher gravitational potential by an aircraft, was compared by short pulses of laser light with the rate of similar clocks on the ground. The plane speed was such as to make as small as possible the special-relativistic effect on the measured time, i.e. it is possible to get small the terms in β^2 in such measurements.

It is worth stressing that the energy-dependent gravitational metric we are looking for is to be regarded as a local representation of gravitation, because the experiments considered took place in a neighbourhood of Earth, and therefore at a small scale with respect to the planetary or galactic ranges of gravity.

Alley and co-workers verified to about 1.5% the validity of the time dilatation formula for a moving clock derived from the Einsteinian gravitational metric, see Fig. 1.3, i.e.

$$\Delta\tau = \Delta t\sqrt{1 + \frac{2G}{c^2} - \frac{v^2}{c^2}}, \tag{1.9}$$

where v is the clock speed and G the Newtonian gravitational potential. They, therefore, utilized the data for a line measurement. On the contrary, the data will be used here for a spectrum analysis, due to the fact that one is interested in deriving the expression of the time metric parameter b_0 as a function of the gravitational energy E.

The difference in elapsed proper time between the clock on the aircraft and the clock on the ground is

$$\Delta T = \sqrt{g_{00}}\Delta\tau - \Delta\tau = \Delta\tau(\sqrt{g_{00}} - 1), \tag{1.10}$$

or, on account of the definition $g_{00} = b_0^2$:

$$\Delta T = \Delta\tau(b_0 - 1). \tag{1.11}$$

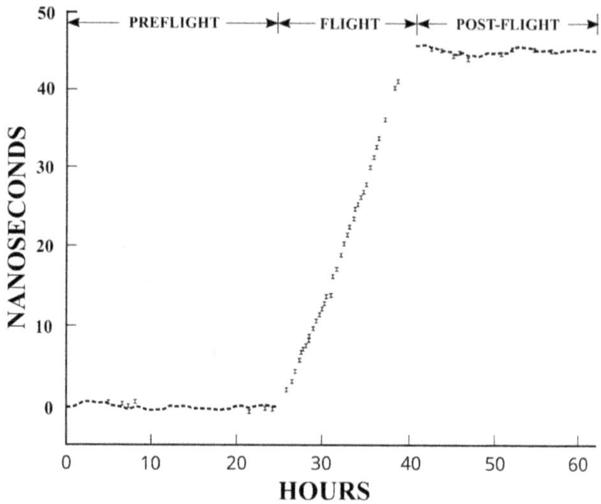

Fig. 1.3. Relative rate ΔT, of the flying clock and the clock at rest in function of time.

By taking, for the test time parameter, the form

$$b_0(E) = \left[1 + \left(\frac{E}{E_0}\right)^n\right] \tag{1.12}$$

with the factor 1 following from the requirement $\Delta T = 0$ for $E = 0$,
Equation (1.11) becomes

$$\frac{\Delta T}{\Delta \tau} = \left(\frac{E}{E_0}\right)^n. \tag{1.13}$$

Moreover, it is reasonable to assume

$$E \equiv E(q), \tag{1.14}$$

where q is the aircraft quote over Earth surface in Alley's experiment. Since it must be $E = 0$ for $q = 0$, it can be put

$$E(q) \propto q^k. \tag{1.15}$$

The simplest assumption is postulating for the energy the form

$$E = \varepsilon \left(\frac{g}{c^2}\right) q, \tag{1.16}$$

where now g is the gravity acceleration, for q in feet and ε in eV, $\frac{g}{c^2} = 3.324 \times 10^{-17}$ (feet)$^{-1}$ then ε is an energy characteristic of the system considered. This agrees with the standard form of the energy of a body of mass m at height q in the Earth gravitational field, $E = mgq$, if we identify ε with the rest energy of the body, $\varepsilon = mc^2$. The most natural choice for ε (which, in our case, represents a characteristic energy of the time measuring device) is the energy gap $h\nu$ of the Cs^{133} hyperfine transition ($h = 4.136 \times 10^{-15}\,\mathrm{eV} \cdot \mathrm{s}$; $\nu = 9.193 \times 10^{9}\mathrm{s}^{-1}$), because it provides the time standard, which triggers the time interval measurements, amplified by the atomic clock device. So, Eq. (1.16) becomes

$$E = h\frac{v}{c^2}gq. \tag{1.17}$$

Let us notice that, in principle, both the frequency ν and the speed c may 'a priori' depend on the quote q of the flying clock. For the frequency, the reason is very easily seen because, on account of Eq. (1.9), it is

$$v(q) = v_0 g_{00}^{-1/2} = v_0 b_0^{-1}. \tag{1.18}$$

However, such a dependence on the altitude does not affect time measurements provided that the atomic transition occurs at a fixed quote. This is just the case of the Alley experiment, where the proper time measurements were taken with the aircraft flying at a given altitude. The dependence of c on q is more subtle. It is due to the fact that, according to the formalism of deformed space time, c in Eqs. (1.16 and 1.17) is no longer the speed of light which is only considered the maximal causal speed for electromagnetic interactions [1,5], but it is to be regarded as the maximal causal speed u_{grav} of gravitation, and therefore expressed by

$$c(q) = u_{k_{\mathrm{grav}}} = \frac{b_0}{b_k}c. \tag{1.19}$$

However, in this case, too, the dependence of the speed on q can be neglected. This can be seen as follows. Firstly on physical grounds and by analogy with the case of the electromagnetic and the leptonic metrics which exhibit the same functional dependence on the energy, see next Section 1.3, it can be assumed that the energy-dependent metric for gravitation behaves as the hadronic metric, see next Section 1.4. This implies that we have $b_1 = \text{const}$, $b_2 = \text{const}(b_1 \neq b_2)$, $b_3(E) = b_0(E)$, so that by putting the constants equal to 1, without any loss of generality

$$u_{1_{\text{grav}}} = u_{2_{\text{grav}}} = cb_0; \quad u_{3_{\text{grav}}} = c. \tag{1.20}$$

In the hypothesis that the change of energy, one is interested into, is along the $k = 3$ direction, it is possible therefore to set

$$u_{3_{\text{grav}}} = c(q) = c. \tag{1.21}$$

Let us notice, however, that in both cases the q-dependence of the frequency and the light speed can actually be neglected. Indeed, the change in the expression (1.17) of the energy due to either effect is:

$$E = \frac{h\nu_0}{c^2} gq \left(1 + \frac{g}{c^2} q\right)^{-3/2} \simeq \frac{h\nu_0}{c^2} gq \tag{1.22}$$

(being $\frac{g}{c^2} = 3.324 \cdot 10^{-17}$ feet^{-1}).

Therefore, we can, in full generality, assume for E the form (1.22) with c the light speed in vacuum and $\nu = \nu_0$. Thus, Eq. (1.13) takes the form

$$\frac{\Delta T_{\text{grav}}}{\Delta \tau} = \left[h\nu_0 \left(\frac{g}{c^2}\right) \left(\frac{1}{E_0}\right) q \right]^n, \tag{1.23}$$

where $\Delta\tau = 1\,\text{day} = 8.640 \cdot 10^{13}$ ns, and the suffix 'grav' means that the time dilation is due only to gravitational effects.

Actually, the experimental data on ΔT do contain a term of kinetic origin ΔT_{kin}, which must be subtracted out according to the formula:

$$\Delta T_{\text{grav}} = \Delta T + \Delta T_{\text{kin}}, \tag{1.24}$$

with

$$\Delta T_{\text{kin}} = \left(\frac{v_0}{c}\right)^2 \frac{\Delta T}{2}. \tag{1.25}$$

Let us notice that the kinetic correction does still have the standard expression, since the e.m. energy-dependent metric reduces to the Minkowskian one for the energy range of the Alley experiment, $E > E_{0_{em}} = 4.5\,\mu eV$ (see Section 1.1).

Now we can rewrite Eq. (1.23) as

$$\frac{\Delta T_{\text{grav}}}{\Delta \tau} = A^n q^n, \tag{1.26}$$

with

$$A = h_0 \left(\frac{g}{c^2}\right)\left(\frac{1}{E_0}\right). \tag{1.27}$$

Putting $y = \frac{\Delta T_{\text{grav}}}{\Delta \tau}$, $x = q$, we get therefore the fit function

$$y = A^n x^n. \tag{1.28}$$

The values of the parameters n, A obtained by the fit are [7]

$$\begin{aligned} n &= 0.9375 \pm 0.0047, \\ A^n &= (6.250 \pm 0.133) \cdot 10^{-17} (feet)^{-n}, \end{aligned} \tag{1.29}$$

with confidence parameter $R^2 = 0.9519$, these fit parameters are updated with respect to those ones reported in [1] and give results more confident.

Let us discuss the physical consequences of the above fit.

According to the results of the fit, it can be concluded that the energy-dependent time coefficient of the gravitational metric reads according Eq. (1.12)

$$b_0(E) = 1 + \frac{E}{E_0}, \tag{1.30}$$

where the gravitational energy threshold E_0 is given by

$$E_0 = E_{0,\text{grav}} = \frac{1}{A} h v_0 \left(\frac{g}{c^2}\right), \tag{1.31}$$

with the value of A obtained by the fit, one has

$$E_{0,\text{grav}} = 20.2 \pm 0.1\,\mu eV. \tag{1.32}$$

For further development it is useful to note that $E_{0,\text{grav}}$ is approximately of the same order of magnitude of the thermal energy corresponding to the 2.7° K cosmic background radiation in the Universe.

The time-dilation formula corresponding to the time coefficient (1.30) reads now

$$d\tau = \left(1 + \frac{E}{E_0}\right) dt \tag{1.33}$$

different from that one deduced by the Einsteinian gravitational metric, which is:

$$d\tau = \left(1 + \frac{E}{E_0}\right)^{1/2} dt. \tag{1.34}$$

As to the explicit form of the spatial part of the gravitational metric, on which the experimental data do not provide any information, two possibilities are open:

(1) *The spatial, 3-dimensional metric is Euclidean, i.e.* $b_1 = b_2 = b_3 = 1$; it is therefore

$$g_{\text{grav}}(E) = \text{diag}\left[\left(1 + \frac{E}{E_{0,\text{grav}}}\right)^2, -1, -1, -1\right]. \tag{1.35}$$

(2) *The spatial metric is anisotropic and energy-dependent, i.e.* the 4-dimensional metric has a structure similar to the hadronic one (see Section 1.4), namely,

$$g_{\text{grav}}(E) = \text{diag}\left[\left(1 + \frac{E}{E_{0,\text{grav}}}\right)^2, -b_1^2, -b_2^2, -\left(1 + \frac{E}{E_{0,\text{grav}}}\right)^2\right], \tag{1.36}$$

with, in general, $b_1^2 \neq b_2^2$.

Now, we note and stress that the gravitational metric is *over-Minkowskian*, since it reaches the flat space time, the Minkowskian limit, from above for energies $E > E_{0,\text{grav}}$.

1.3 Decay of K_s^0 Experiments

Let us now discuss the leptonic interaction. In order to derive the phenomenological leptonic metric, a purely leptonic decay has been analyzed, that of the meson K_s^0, whose meanlife is experimentally known in a wide energy range ($30 \div 350\,\text{GeV}$) [8, 9]. This is a useful case, motivated by the fact that the K_s^0 decay violate charge conjugation-parity (CP) invariance, and therefore due to the *CPT* theorem, it should provide indirect evidence for violation of time reversal. Now, as in the electromagnetic case, it is assumed a spatially isotropic metric. It will be seen in the following that the isochronism with the usual Minkowski metric, i.e. $b_0^2 = 1$, is derived from the phenomenological analysis, without any need of postulating it.

In this case, we exploit the law of time dilation in deformed space time using the following energy dependent form:

$$\Delta t = \frac{\Delta t_0}{\left[1 - \left(\frac{b}{b_0}\right)^2 + \left(\frac{b}{b_0}\right)^2 \left(\frac{m_0}{E}\right)^2\right]^{1/2}}. \tag{1.37}$$

Although the known data on K_s^0 lifetime belong to two different measurements, taken in two different experiments [8, 9], they had been made statistically homogeneous, thus allowing one to treat them as a single set of data in the analysis by relation (1.37). The main steps in deriving the leptonic metric are: firstly, the data interpolation via the Einsteinian–Minkowskian linear law is compared with the interpolation via Eq. (1.37), thus providing the conclusion that this law does not yield results consistent with a constant metric, see Figs. 1.5(a) and 1.5(b); secondly, one carries out, still via Eq. (1.37), an interpolation of the experimental data in subintervals at different energies, so finding the values of b^2 and b_0^2 at various energies, see Figs. 1.6(a) and 1.6(b). Such a procedure allows one to conclude that, at least in the range considered, b_0^2 does not vary with energy unlike b^2 which changes with energy, see Figs. 1.6(a) and 1.6(b). The interpolation of the values of parameters b^2 and b_0^2 permits then to derive an explicit expression of the metric, which yields the phenomenological description of leptonic interaction, regarded as an interaction not derivable from a potential [10, 11].

The results of the fit yield the following metric:

$$
\begin{aligned}
g_{\text{lep}}(E) &= \text{diag}(b^2_{0,\text{lep}}(E), -b^2_{\text{lep}}(E), -b^2_{\text{lep}}(E), -b^2_{\text{lep}}(E)),\\
&\quad b^2_{0,\text{lep}}(E) = 1\\
b^2_{\text{lep}}(E) &= \begin{cases} (E/E_{0,\text{lep}})^{1/3}, & 0 \le E \le E_{0,\text{lep}} = 80.4 \pm 0.2\,\text{GeV} \\ 1, & E_{0,\text{lep}} < E \end{cases}.
\end{aligned}
\tag{1.38}
$$

The meaning of $E_{0,\text{lep}}$ is again the energy value at which the metric becomes Minkowskian, it is worth stressing that such a value of the energy corresponds to the mass of the W-boson, through which the K^0_s decay does occur. Moreover, $g_{\text{lep}}(E)$ is *sub-Minkowskian*, like the electromagnetic metric, namely it reaches the flat space time, the Minkowskian limit, from below for energies $E < E_{0,\text{lep}}$. We therefore recover the well known fact that, within the unified model of electromagnetic and leptonic interactions, $E_{0,\text{lep}}$ is the energy scale at which the leptonic and electromagnetic interactions are mixed, or rather, according our geometrical view, reach the same metric which is Minkowskian, see Fig. 1.4.

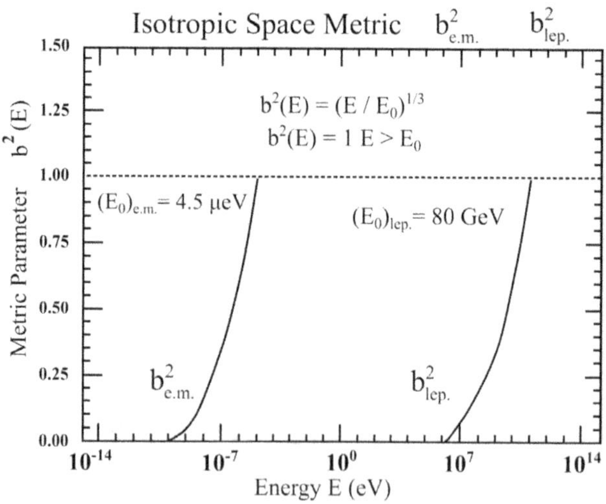

Fig. 1.4. Plot of the behaviour of the electromagnetic and leptonic spatial parameters, in the hypothesis of a spatially isotropic metric.

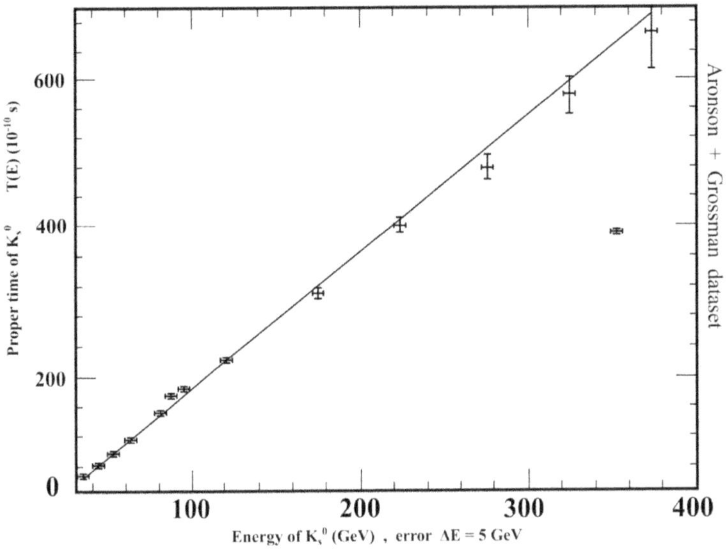

Fig. 1.5(a). Fit of K^0_S lifetime by Einsteinian–Minkowskian linear function.

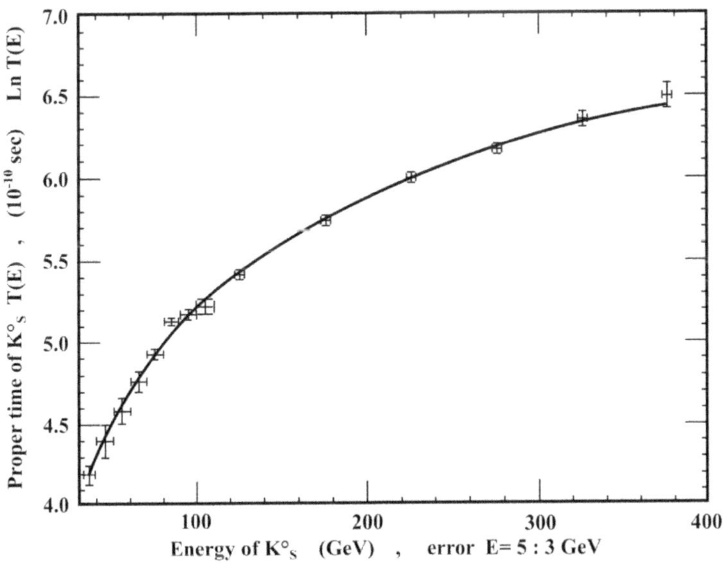

Fig. 1.5(b). Fit of K°$_S$ lifetime by nonlinear function Eq. (1.37).

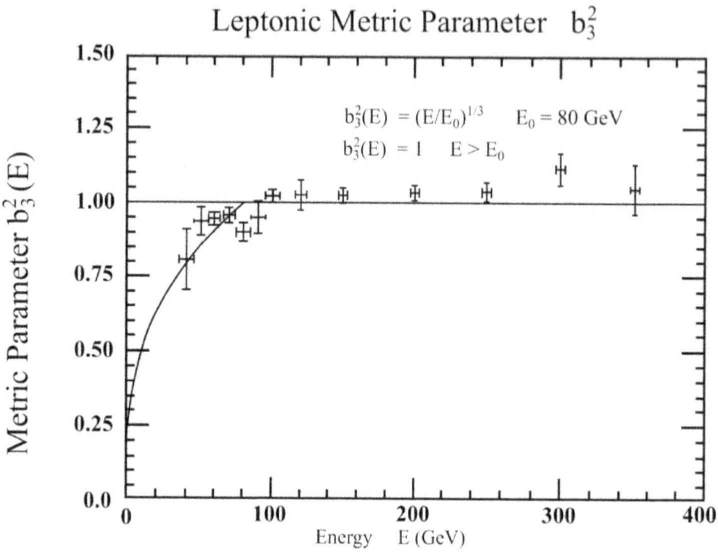

Fig. 1.6(a). Leptonic metric parameter b^2 vs. energy.

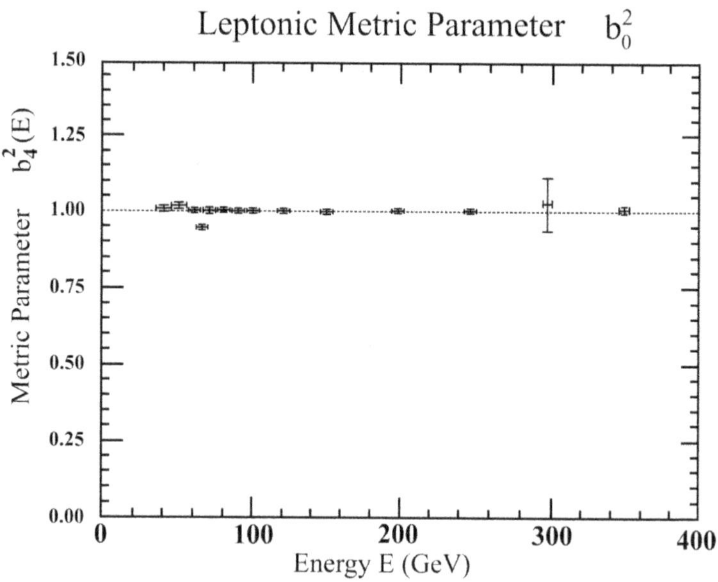

Fig. 1.6(b). Leptonic metric parameter b_0^2 vs. energy, showing the isochronism of the leptonic metric with the usual Minkowskian one.

The new interpretation provided by the formalism of deformed space time is that $E_{0,\text{leptonic}}$ is the energy scale at which the leptonic and electromagnetic metrics measure space-time separations in the same way.

Therefore, it is not surprising that, using the same interpolation conditions, the leptonic metric (1.38) has the same form of the e.m. metric (1.8).

Finally, Fig. 1.4 shows in an energy logarithmic scale the spatial parameters of the spatial isotropic electromagnetic metric (1.8) and of the spatial isotropic leptonic metric (1.38). Notice that, due to the energy scale used, it cannot be stated that either parameter vanishes at zero energy, at least at the present status of development of the Deformed Space Time Theory [5].

1.4 Proton Antiproton Experiment

In order to derive the expression of the hadronic metric, we consider the anisotropic correlation function for the Bose–Einstein (BE) correlation occurring for mesons (bosons) as last stage of the proton antiproton annihilation: $C_{(2)}^{\widetilde{AN}}$.

This anisotropic correlation function can be expressed in terms of the transverse momentum transfer q_t as:

$$C_{(2)}^{\widetilde{AN}} = 1 + b_0^2 \left[e^{-q_t^2} - b_1'^2 e^{-q_t^2/b_1'^2} - b_2'^2 e^{-q_t^2/b_2'^2} - b_3'^2 e^{-q_t^2/b_3'^2} \right]$$
$$b_k' = \tfrac{b_k}{b_0}, \quad k = 1, 2, 3. \tag{1.39}$$

The products have been carried out classically in the usual Minkowski space, since the definition of q_t is the experimental one, or rather we do not use the definition of the scalar product in the deformed space time [1].

Equation (1.39) is then used to interpolating the data concerning the mesons produced in proton-antiproton annihilation at total energy given by the measurements taken by the UA1 detector at CERN in the "ramping run" in 1984 [12]. The annihilation has been chosen in order to have the same energy in the center-of-mass and in the laboratory frames, so as to refer the values of the four parameters b_μ^2 to a unique energy value.

The results thus obtained for the b_k^2's from Eq. (1.39) show that the phenomenon of BE correlation is indeed spatially anisotropic, so that the related metric is anisotropic. Therefore, we can state that spatial anisotropy is, in general, a peculiar feature of hadronic interactions. In spite of the phenomenological nature of the correlation function (1.39), it allows one to get new results, like the loss of axial symmetry of the BE phenomenon, the geometrical description, via the parameters b_k, of the annihilation region, and the q_t value at which the correlation effect begins [13].

This analysis gives a strong evidence that hadronic interaction can be treated neither in terms of a metric with constant parameters, nor in terms of a flat Minkowski metric.

By interpolating the experimental data, there follows that the explicit form of the metric g_{had} which describes the hadronic interaction, as a function of $E = \sqrt{s}$, where s is the Mandelstam invariant, reads as:

$$\begin{aligned} g_{\text{had}}(E) &= \text{diag}(b_{0,\text{had}}^2(E), -b_{1,\text{had}}^2, -b_{2,\text{had}}^2, -b_{3,\text{had}}^2(E)) \\ b_{0,\text{had}}^2(E) &= b_{3,\text{had}}^2(E) \\ &= \begin{cases} 1, & 0 \le E \le E_{0,\text{had}} = 367.5 \pm 0.4\,\text{GeV} \\ (E/E_{0,\text{had}})^2 & E_{0,\text{had}} < E \end{cases} \\ b_{1,\text{had}}^2 &= (\sqrt{2}/5)^2 \\ b_{2,\text{had}}^2 &= (2/5)^2. \end{aligned} \tag{1.40}$$

In this case, too, $E_{0,\text{had}}$ is to be meant as the energy value at which the metric becomes Minkowskian.

The results of the fit giving behaviour of the metric parameters are shown in Fig. 1.7 where there is the ratio between the time parameter $b_{0,\text{had}}^2$ and one spatial parameter, we choose $b_{3,\text{had}}^2$. Moreover, for $E < E_{0,\text{had}}$, the hadronic parameters $b_{0,\text{had}}^2$ and $b_{3,\text{had}}^2$, take the Minkowskian value, which, for energy range $80.4\,\text{GeV} < E < 367.5\,\text{GeV}$, coincides with that of both e.m. and leptonic interactions, Fig. 1.7 shows the behaviour of the ratio $b_{0,\text{had}}^2(E)/b_{3,\text{had}}^2(E)$, whence it is clearly seen that we can put $b_3^2(E) = b_0^2(E)$, see Eq. 1.40. In Fig. 1.8 are reported all the hadronic metric parameters b_μ coming from the results of the fits, they are plotted together in function of energy, in order to emphasize the spatial anisotropy of the metric.

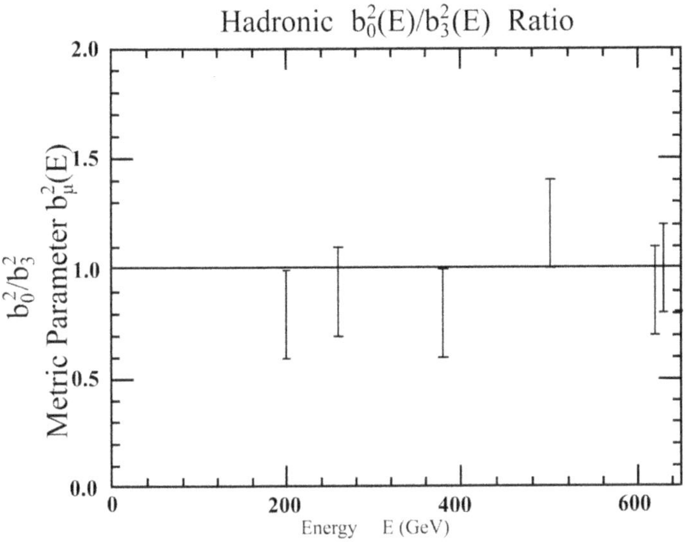

Fig. 1.7. Ratio of the hadronic metric parameters b_0^2/b_3^2.

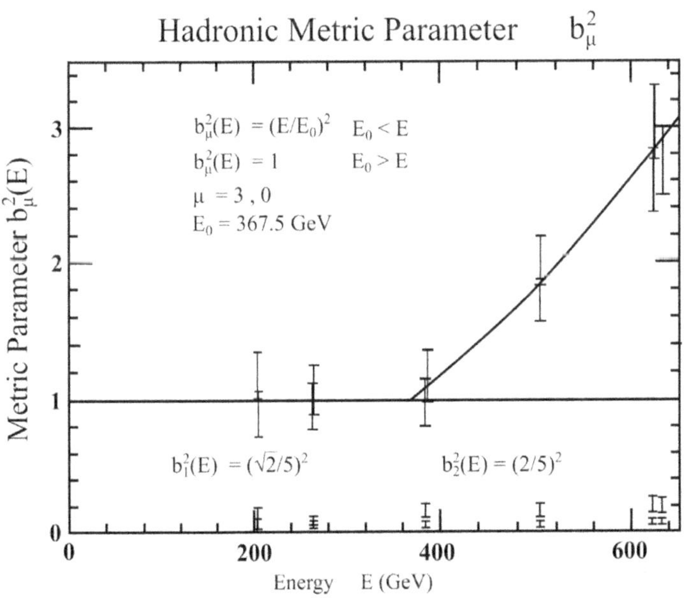

Fig. 1.8. Behaviour with energy of all four hadronic metric parameters b_μ^2.

Two spatial parameters are constant, but different in value, whereas the third one varies and is over Minkowskian as energy increases or rather for energies $E > E_{0,\text{had}}$. Let us recall that the metric parameters are related to the space sizes and to the meanlife a_0 of the BE 'fireball', which is the space time region and volume where the proton antiproton annihilation occurs. In particular, by using the minimum bias of UA1 with energy $\sqrt{s} = 630\,\text{GeV}$, one gets for a_0 the value

$$a_0 = \tau = (1.09 \pm 0.01) \times 10^{-24} s. \tag{1.41}$$

The corresponding energy 'width' Γ of the fireball, given by

$$\Gamma \equiv \frac{\hbar}{a_0} = (0.61 \pm 0.02)\,\text{GeV} \tag{1.42}$$

is consistent with the experimental data, if one interprets Γ as that value of q_t such that $C_{(2)}^{\text{exp}} \lesseqgtr 1$ *if* $q_t \gtreqless \Gamma$ where $C_{(2)}^{\text{exp}}$ is the classical exponential form of the isotropic correlation function for the BE correlation.

We will come back and use these results in the next when we are going to explore the space time description of the energy coming from proton antiproton annihilation using the concept of Deformed Space Time theory but then including the problem of recovering lost symmetry. In Fig. 1.9, we show, for a comparison, the spatial parameters $b_{3,\text{lep}}^2$ (isotropic case) and $b_{3,\text{had}}^2$ of the leptonic metric see Eq. (1.38) and of the hadronic one see Eq. (1.40), respectively. As already noted we see clearly that the leptonic parameter exhibits a sub-Minkowskian behaviour, unlike the hadronic one, which is over-Minkowskian. Moreover, both parameters take unity value in the range $80.4 \div 367.5\,\text{GeV}$, like in the e.m. case.

Quite likely, the fact that in the $80.4 \div 367.5\,\text{GeV}$ energy range the metric looks Minkowskian for electromagnetic, leptonic and hadronic interactions, may explain why perturbation methods are so successful in describing both leptonic and hadronic reactions, for such intervals of energy of the reaction final products. This apparent success occurs despite the fact that these methods would be *a priori* expected to fail, like in perturbative quantum chromodynamics. Let us explicitly notice that the hadronic metric (1.40) is not always isochronous

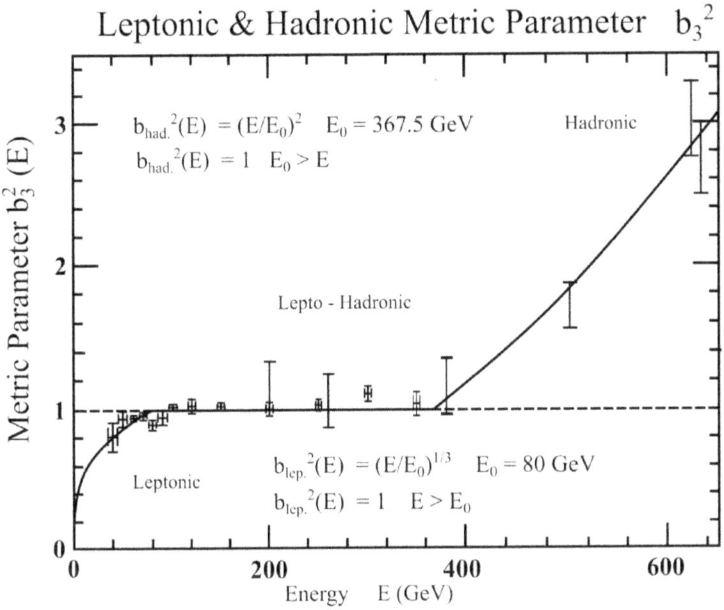

Fig. 1.9. Comparing the spatial parameters b_3^2 for leptonic and hadronic metrics.

with the usual Minkowski metric ($b_0^2 = 1$) which characterizes the electromagnetic interaction. Actually, it follows from (1.40) that it is $b_{0,\text{had}}^2 \neq 1$ for $E_{0,\text{had}} < E$. Such a case is not new; indeed, as is well known, the same happens for the gravitational interaction, as shown e.g. by the various measurements of red or blue shifts of electromagnetic radiation in a gravitational field, or by the relative delays of atomic clocks put at different heights in presence of gravity as used in Section 1.3 for gravitational energy metric.

Now it is worth noticing that, by analogy with the physical meaning attributed to $E_{0,\text{lep}}$ as the energy scale of the intermediate vector bosons for electroweak interactions, an analogous interpretation can be given to $E_{0,\text{had}}$. A possible suggestion is that the value of $E_{0,\text{had}}$ does represent the energy scale corresponding to the upper limit of the mass of the Higgs boson, that breaks the gauge invariance of the lepto-hadronic mixed interactions by endowing the leptonic interaction carriers with mass. On the other hand, in the framework of the Deformed Space Time theory, it is possible to give inertial mass to stable particles on the basis of Lorentz invariance breakdown parameter and the energy threshold E_0 of a related interaction: this is the concept of 'geometrical' mass of a particle [1]. Therefore, such an interpretation of $E_{0,\text{had}}$ requires a deep reflection on the possible links between gauge invariance and Lorentz invariance breakdowns, which, in our opinion, can be only dealt with in the framework of a five-dimensional formulation, which we will consider in the next.

Let us investigate the possible implications of such an anisochronism of the hadronic metric. We denote by dt_{had} the time interval taken by a certain hadronic process for a particle at rest, we can call it the 'hadronic clock'. The same process, when referred to a Minkowskian electromagnetic metric, will take a time $dt_{\text{e.m.}}$ to happen. Borrowing methods and notation from the general theory of relativity, we can state that, for a particle at rest:

$$\frac{dt_{\text{had}}}{dt_{\text{e.m.}}} = \frac{1}{\sqrt{g_{00,\text{had}}}}. \tag{1.43}$$

Since for the deformed metric (1.40) it is $g_{00,\text{had}} = b^2_{0,\text{had}}(E)$, we get

$$\frac{dt_{\text{had}}}{dt_{\text{e.m.}}} = \begin{cases} 1, & 0 \leq E \leq E_{0,\text{had}} = 367.5 \pm 0.4\,\text{GeV} \\ E_{0,\text{had}}/E, & E_{0,\text{had}} < E \end{cases}. \tag{1.44}$$

Equation (1.44) yields the 'law of time deformation' for hadronic interaction.

Figure 1.10 shows the behaviour of law (1.44), i.e. the plot of dt_{had} vs. energy in units of $dt_{\text{e.m.}}$. It is easily seen that there is isochronism at low energies, the physical processes have the same time rate either when referred to a hadronic metric or to an electromagnetic one, whereas there is a time contraction at high energies. In other words, hadronic processes are faster when observed with respect to an electromagnetic metric.

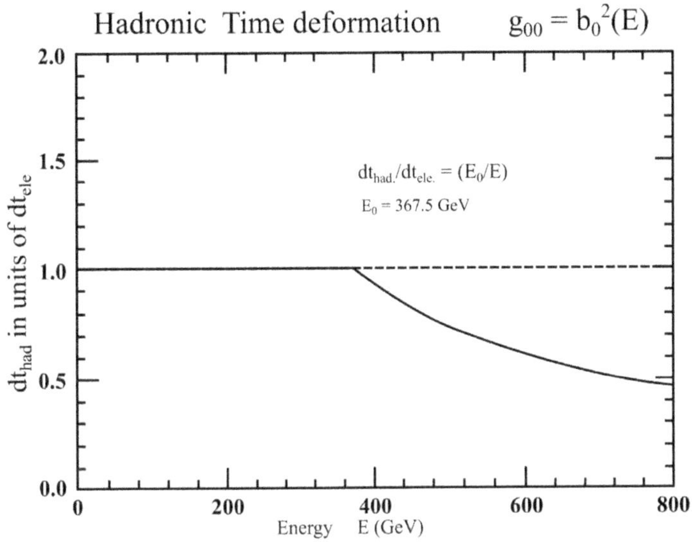

Fig. 1.10. Time deformation law in a hadronic field vs. the energy.

1.5 Review of Phenomenological Metrics

The main aim of the present Paragraph is just to provide a first review of the phenomenological metrics derived for all four fundamental interactions, by stressing their general structure and their features on a physical and phenomenological ground. For the sake of completeness we repeat this list in the next chapter where we will be focused on the mathematical details as a premise to further developments. The local breakdown of Lorentz invariance may be envisaged for all four fundamental interactions, electromagnetic, gravitational, leptonic and hadronic, whereby one gets evidence for a departure of the space time metric from the Minkowskian one, at least in the energy range examined in the reported experiments. Let us review the basic features of the energy dependent phenomenological metrics derived above:

(1) Both the electromagnetic and the leptonic metric show the same functional behaviour, namely

$$g(E) = \text{diag}(1, -b^2(E), -b^2(E), -b^2(E)), \tag{1.45}$$

$$b^2(E) = \begin{cases} \left(\frac{E}{E_0}\right)^{1/3}, & 0 < E \leq E_0 \\ 1, & E_0 < E \end{cases} \tag{1.46}$$

$$1 + H(E_0 - E)\left[\left(\frac{E}{E_0}\right)^{1/3} - 1\right] E > 0, \tag{1.47}$$

where H is the Heaviside step function, with the only difference between them being the threshold energy E_0, i.e. the energy value at which the metric parameters are constant, namely the metric becomes Minkowskian: $g_{\mu\nu\text{int}}(E \geq E_0) \equiv g_{\mu\nu\text{flat}} = \text{diag}(1, -1, -1, -1)$; the fits to the experimental data yield

$$E_{0,\text{e.m}} = (4.5 \pm 0.2)\,\mu eV; \quad E_{0,\text{lep}} = (80.4 \pm 0.2)\,GeV. \tag{1.48}$$

Note that for either interaction the metric is isochronous, spatially isotropic and sub-Minkowskian, it approaches the Minkowskian limit from below (for $E < E_0$). Both metrics are therefore Minkowskian for $E > E_{0\text{lep}} \simeq 80\,GeV$, and then this formalism is fully consistent with electroweak unification, which occurs at an energy scale $\sim 100\,GeV$.

(2) The hadronic metric reads

$$g_{\text{had}}(E) = \text{diag}(b^2_{0,\text{had}}(E), -b^2_{1,\text{had}}(E), -b^2_{2,\text{had}}(E), -b^2_{3,\text{had}}(E)). \tag{1.49}$$

$$b^2_{1,\text{had}}(E) = \left(\frac{\sqrt{2}}{5}\right)^2; \quad b^2_{2,\text{had}}(E) = \left(\frac{2}{5}\right)^2, \forall E > 0. \tag{1.50}$$

$$b^2_{0,\text{had}}(E) = b^2_{3,\text{had}}(E)$$
$$= \begin{cases} 1, & 0 < E \leq E_{0,\text{had}} \\ \left(\frac{E}{E_{0,\text{had}}}\right)^2, & E_{0,\text{had}} < E \\ 1 + H(E - E_{0,\text{had}})\left[\left(\frac{E}{E_{0,\text{had}}}\right)^2 - 1\right], & E > 0 \end{cases}. \tag{1.51}$$

with

$$E_{0,\text{had}} = (367.5 \pm 0.4)\,\text{GeV}. \tag{1.52}$$

In this case, contrarily to the electromagnetic and the leptonic ones, a deformation of the time coordinate occurs; moreover, the three-space is anisotropic, with two spatial parameters constant, but different in value, and the third one variable with energy in the same way as the time one.

(3) The explicit form of the gravitational energy-dependent metric is

$$g_{\text{grav}}(E) = \text{diag}(b_{0,\text{grav}}^2(E), -b_{1,\text{grav}}^2(E), -b_{2,\text{grav}}^2(E), -b_{3,\text{grav}}^2(E)). \tag{1.53}$$

$$b_{0,\text{grav}}^2(E) = b_{3,\text{grav}}^2(E)$$
$$= \begin{cases} 1, & 0 < E \leq E_{0,\text{grav}} \\ \frac{1}{4}\left(1 + \frac{E}{E_{0,\text{grav}}}\right)^2, & E_{0,\text{grav}} < E \\ 1 + H(E - E_{0,\text{grav}}) & \\ \quad \times \left[\frac{1}{4}\left(1 + \frac{E}{E_{0,\text{grav}}}\right) - 1\right], & E > 0 \end{cases} \tag{1.54}$$

(the coefficients $b_{1,\text{grav}}^2(E)$ and $b_{2,\text{grav}}^2(E)$ are presently undetermined at phenomenological level), with

$$E_{0,\text{grav}} = (20.2 \pm 0.1)\,\mu eV. \tag{1.55}$$

It is worth stressing the analogy between the hadronic and the gravitational metrics. In both cases, a deformation of the time coordinate occurs. Moreover, one of the spatial parameters, we conventionally assumed as the third one, varies with energy as does the time one in an over-Minkowskian way, namely, they approach the Minkowskian limit from above ($E_0 < E$). The other two spatial parameters are constant, but different in value for the hadronic case, the three-space is anisotropic for hadronic interaction, whereas nothing can be said for the gravitational case on an experimental basis.

References

[1] F. Cardone and R. Mignani *Energy and Geometry.* Ed. World Scientific 2004.
[2] A. Enders and G. Nimtz *J. Phys I (France)*, 2, 1693, 1992.
[3] A. Enders and G. Nimtz *Phys. Rev. E*, 48, 632, 1993.
[4] W. Heitmann and G. Nimtz *Phys. Lett. A*, 154, 1994.
[5] F. Cardone and R. Mignani *Deformed Space Time* Ed. Springer 2007.
[6] C. O. Alley. *Relativity and Clocks.* In: Proceeding of the 33rd Annual Symposium on Frequency Control, Elect. Ind. Ass. Washington DC 1979.
[7] F. Cardone, R. Mignani *Int. J.Modern Phys. A*, 14, 3799, 1999.
[8] S. H. Aronson *et al. Phys. Rev. D* 28, 632, 1983.
[9] N. Grossman *et al. Phys. Rev. Lett.*, 59, 18, 1987.
[10] F. Cardone *et al. J. Phys. G* 18, L61 and l41, 1992.
[11] F. Cardone, R. Mignani *Found. of Phys.*, 29, 1735, 1999.
[12] UA1 Collaboration. *Phys. Lett. B*, 226, 410, 1989.
[13] F. Cardone and R. Mignani. *JETP,* 83, 435, (*Zh. Eksp. Teor. Fiz.* 110, 793), 1996.

Chapter 2

Four Dimensional Energy Metrics

We resume now the energy metrics of the fundamental interactions, electromagnetic, gravitational, leptonic and hadronic obtained from the analysis of experimental data. We note that these metrics have a common mathematical structure, since the metric elements are function of a power of energy E. We will also see that they have a sort of 'coupling' between a sub-Minkowskian and an over-Minkowskian giving rise to the couples electromagnetic-gravitational and leptonic-hadronic.

Another feature is the 'natural' order among the energy thresholds $E_{0\text{int}}$ of the metrics, they are in growing order of their value from electromagnetic to hadronic, moreover they are in same proportion where the ratio of each couple of $E_{0\text{int}}$ (over-Minkowskian)/$E_{0\text{int}}$ (sub-Minkowskian) is constant when considering the 'coupling' electromagnetic-gravitational and leptonic–hadronic. These evidences give us the possibility to set 'recursive rules' for energy thresholds.

2.1 Electromagnetics Metric

The experiments considered were those on the superluminal propagation of e.m. waves in conducting waveguides with variable section [1]. The introduction, in this framework, of a deformed space time was motivated by ascribing the superluminal speed of the signals to some nonlocal e.m. effect, inside the narrower part of the waveguide, which can be described in terms of an effective deformation of space time inside the barrier region. Since one is dealing with electromagnetic forces, which are usually described by the Minkowskian metric, it was possible to assume $b_0^2 = 1$ and this is also justified by the fact that all the relevant deformed quantities depend actually on the ratio b/b_0. Assuming moreover an isotropically deformed three-space ($b_1 = b_2 = b_3 = b$), the deformed e.m. energy dependent metric results in the following expression where D stands for 'deformed' and where H is the Heaviside step function:

$$g_{D,\text{e.m.}}(E) = \text{diag}(1, -b^2_{\text{e.m.}}(E), -b^2_{\text{e.m.}}(E), -b^2_{\text{e.m.}}(E)), \tag{2.1}$$

$$\begin{aligned} b^2_{\text{e.m.}}(E) &= \begin{cases} (E/E_{0\text{e.m.}})^{1/3}, & 0 \leq E < E_{0\text{e.m.}} \\ 1, & E_{0\text{e.m.}} < E \end{cases} \\ &= 1 + H(E_{0\text{e.m.}} - E)\left[\left(\frac{E}{E_{0\text{e.m.}}}\right)^{1/3} - 1\right], \quad E > 0. \end{aligned} \tag{2.2}$$

The threshold energy $E_{0\text{e.m.}}$ is the energy value at which the metric parameters attain a constant value, and the metric becomes flat Minkowskian. The phenomenological value obtained is

$$E_{0,\text{e.m.}} = (4.5 \pm 0.2)\, \mu eV. \tag{2.3}$$

Note that the value obtained for $E_{0\text{e.m.}}$ is of the order of the energy E_{coh} corresponding to the coherence length of a photon for radio-optical waves used in waveguide experiments. Indeed, for such waves it is $\nu \approx 100\, MHz$, corresponding to 10^{-8} s, and therefore $\lambda \approx 10^2$ cm so that

$$E_{\text{coh}} \equiv \frac{hc}{\lambda} \simeq 1\, \mu eV. \tag{2.4}$$

It must be stressed that the departure of the metric from the Minkowskian one is a signature of the Lorentz invariance (LI) breakdown in its usual meaning. Therefore breaking LI in electromagnetic processes is expected to occur at low energies and not high, as commonly thought. The deformed electromagnetic metric $g_{D,\text{e.m.}}$ is sub-Minkowskian reaching the flat Minkowskian metric for energy values increasing up to $E_{0\text{e.m.}}$, as we will see it is the first metric of the first couple of metrics together with the gravitational one.

2.2 Gravitational Metric

It is possible to show that also the gravitational interaction can be described in terms of an energy dependent metric, whose time coefficient was derived by fitting the experimental results on the relative rates of clocks at different heights in the gravitational field of Earth, obtained by C. O. Alley in 1974–1976 [2]. No information can be derived from the experimental data about the space parameters. The energy dependent gravitational metric was thus obtained in the following form where D stands for 'deformed' and where H is the Heaviside step function:

$$\begin{aligned} g_{D,\text{grav}}(E) = \text{diag}(b_{0,\text{grav}}^2(E), -b_{1,\text{grav}}^2(E), -b_{2,\text{grav}}^2(E), \\ - b_{3,\text{grav}}^2(E)). \end{aligned} \tag{2.5}$$

$$b_{0,\text{grav}}^2(E) = \begin{cases} 1, & 0 \le E < E_{0\text{grav}} \\ \frac{1}{4}(1 + E/E_{0\text{grav}})^2, & E_{0\text{grav}} < E \end{cases}$$

$$= 1 + H(E - E_{0,\text{grav}}) \left[\frac{1}{4} \left(1 + \frac{E}{E_{0,\text{grav}}} \right)^2 - 1 \right], \quad E > 0, \tag{2.6}$$

with the coefficients $b_{k,\text{grav}}^2(E)(k = 1, 2, 3)$ undetermined and

$$E_{0,\text{grav}} = (20.2 \pm 0.1)\,\mu eV. \tag{2.7}$$

We note that the value of the threshold energy for the gravitational deformed metric $E_{0,\text{grav}}$ is approximately one order of magnitude less than the thermal energy corresponding to the 2.7 K cosmic microwave background radiation in the Universe ($E_{\text{CMBR}} \approx 232.67\,\mu eV$).

Let us notice that for the gravitational metric, in analogy with the hadronic metric, a deformation of the time coordinate occurs. The time coefficient varies with energy in an over-Minkowskian way, namely, it approaches the flat Minkowskian limit from above ($E_0 < E$). As already noted in Chapter 1 the relation between proper time and coordinate time, corresponding to the time coefficient $b_{0,\text{grav}}^2$

given by (2.6) reads:

$$d\tau = \left(1 + \frac{E}{E_0}\right) dt, \tag{2.8}$$

at variance with that one derived from Einsteinian gravitational metric

$$d\tau = \left(1 + \frac{E}{E_0}\right)^{1/2} dt. \tag{2.9}$$

For the explicit form of the spatial part of the gravitational metric we choose to follow the analogy with the hadronic metric as stated in Chapter 1, therefore the gravitational spatial metric is anisotropic and dependent by energy:

$$\begin{aligned} g_{D,\text{grav}}(E) = \text{diag}(&b^2_{0\text{grav}}(E), -b^2_{1\text{grav}}(E), -b^2_{2\text{grav}}(E), \\ &- b^2_{3\text{grav}}(E). \end{aligned} \tag{2.10}$$

$$\begin{aligned} b^2_{0,3\text{grav}}(E) &= \begin{cases} 1, & 0 \leq E < E_{0\text{grav}} \\ \frac{1}{4}(1 + E/E_{0\text{grav}})^2, & E_{0\text{grav}} < E \end{cases} \\ &= 1 + H(E - E_{0,\text{grav}}) \left[\frac{1}{4}\left(1 + \frac{E}{E_{0,\text{grav}}}\right)^2 - 1\right], \quad E > 0, \end{aligned} \tag{2.11}$$

with, in general, $b^2_1 \neq b^2_2$. Assuming a gravitational deformation of space, allowing also for spatial anisotropy, seems to be the more sound choice on physical ground, on account of the experimental evidences about the speed of gravity [3]–[7]. This is why we shall mean for gravitational metric the form (2.10), (2.11). The deformed metric $g_{D,\text{grav}}$ is over-Minkowskian reaching the flat Minkowskian metric for energy values decreasing to $E_{0\text{grav}}$, as we will see it is the second metric of the first couple of metrics together with the electromagnetic one.

2.3 Leptonic Metric

According to what introduced in Chapter 1 the leptonic metric has been deduced from the experimental input provided by the data on the pure leptonic decay of the meson K_s^0, whose lifetime T is known in a wide energy range 30–350 GeV. We used the deformed law of time dilation as a function of the energy, which reads:

$$T = \frac{T_0}{\left[1 - \left(\frac{b}{b_0}\right)^2 + \left(\frac{b}{b_0}\right)^2 \left(\frac{m}{E}\right)^2\right]^{1/2}}. \tag{2.12}$$

As in the electromagnetic case, an isotropic three dimensional space was assumed, whereas the isochrony with the usual Minkowski metric, or rather $b_0^2 = 1$, has been derived by the fit of (2.12) to the experimental data. The corresponding leptonic metric is given by the following expression where D stands for 'deformed' and H is the Heaviside step function

$$g_{D,\text{lep}}(E) = \text{diag}(b_{0,\text{lep.}}^2(E), -b_{\text{lep.}}^2(E), -b_{\text{lep.}}^2(E), -b_{\text{lep.}}^2(E)), \tag{2.13}$$

$$\begin{aligned} b_{\text{lep.}}^2(E) &= \begin{cases} (E/E_{\text{lep}})^{1/3}, & 0 \leq E < E_{\text{lep}} \\ 1, & E_{\text{lep}} < E \end{cases} \\ &= 1 + H(E_{\text{lep.}} - E)\left[\left(\frac{E}{E_{\text{lep}}}\right)^{1/3} - 1\right], \quad E > 0, \end{aligned} \tag{2.14}$$

with

$$E_{0,\text{lep}} = (80.4 \pm 0.2)\,\text{GeV}. \tag{2.15}$$

The value of $E_{0,\text{lep}}$ is the energy value at which the weak metric becomes Minkowskian, its value corresponds to the mass of the W boson through which the K_s^0 decay occurs. The leptonic metric has the same form of the electromagnetic metric (2.1), (2.2) and the isocrony, obtained by the fit of the deformed time law given by Eq. (2.12), gives $b_{0,\text{lep}}^2 = 1$. Therefore by the Deformed Space Time theory model, one recovers that, at the energy scale $E_{0\text{lep}}$ and over, the leptonic and the electromagnetic interactions are mixed, or rather in geometrical words they measure the distances and the time in the same way independently by the energy of the physical process ruled.

We want now to recall that, in both the electromagnetic and the leptonic case, the metric parameter exhibits a sub-Minkowskian behaviour, where $b(E)$ approaches the Minkowskian value 1 from below ($E < E_0$) as energy increases. So, the deformed leptonic metric $g_{\text{Dlep.}}$ is sub-Minkowskian reaching the flat Minkowskian metric for energy values increasing up to $E_{0\text{lep}}$, as we will see it is the first metric of the second couple of metrics together with the hadronic one.

2.4 Hadronic Metric

The phenomenon considered was the so-called Bose Einstein (BE) effect in the hadronic production of identical bosons in high energy collisions, which consists in an enhancement of their correlation probability. The deformed space time formalism permits to derive a generalized BE correlation function, depending on all the four metric parameters $b_\mu(E)$. By using the experimental data on pion pair production, obtained in 1984 by the UA1 collaboration at CERN studying the products of the proton antiproton annihilation in an energy range of about one order of magnitude, we got the following expression of the hadronic metric:

$$g_{D,\text{had}}(E) = \text{diag}(b_{0,\text{had}}^2(E), -b_{1,\text{had}}^2(E), -b_{2,\text{had}}^2(E), - b_{3\text{had}}^2(E)), \tag{2.16}$$

$$b_{3,0\text{had}}^2(E) = \begin{cases} 1, & 0 \le E < E_{0\text{had}} \\ (E/E_{0\text{had}})^2, & E_{0\text{had}} < E \end{cases}$$

$$= 1 + H(E - E_{0,\text{had}}) \left[\left(\frac{E}{E_{0,\text{had}}} \right)^2 - 1 \right], \quad E > 0, \tag{2.17}$$

$$b_{1,\text{had}}^2(E) = (\sqrt{2}/5)^2; \tag{2.18}$$

$$b_{2,\text{had}}^2 = (2/5)^2, \tag{2.19}$$

where D stands for 'deformed', H is the Heaviside step function and with

$$E_{0,\text{had}} = (367.5 \pm 0.4)\,\text{GeV}. \tag{2.20}$$

The threshold energy $E_{0,\text{had}}$ is still the value at which the metric becomes flat Minkowskian. Let us stress that, in this case, contrarily to the electromagnetic and the leptonic ones, a deformation of the time coordinate occurs. Moreover, the three dimensional space is anisotropic, with two spatial parameters constant, but different in value, and the third one variable with energy in an over-Minkowskian way. It is also worth to recall that the strong metric parameters b_μ admit a physical interpretation: the spatial parameters are related to the spatial sizes of the interaction region, the so called fireball where it is believed the annihilation occurs and pions are produced, whereas the time parameter is essentially the mean life of the process, the fireball duration.

We note that an interpretation can be given to $E_{0,\text{had}}$, which is analogous to the physical meaning which can be attributed to $E_{0,\text{lep}}$ as the energy scale of the intermediate vector bosons for electroweak interactions.

An interpretation is that the value of $E_{0,\text{had}}$ does represent the energy scale corresponding to the upper limit of the mass for the Higgs boson, that is believed to break the gauge invariance of the electroweak mixed interactions by endowing the weak interaction carriers with mass.

Let us explicitly notice that the hadronic metric (2.17), (2.19) is not always isochronous with the usual Minkowski metric ($b^2_{0,\text{had}} \neq 1$). Actually, it follows from (2.17) that it is $b^2_{0,\text{had}} \neq 1$ for $E_{0,\text{had}} < E$. The deformed hadronic metric $g_{D,\text{had}}$ is over-Minkowskian reaching the flat Minkowskian metric for energy values decreasing to $E_{0\text{had}}$, as we will see it is the second metric of the second couple of metrics together with the leptonic one.

2.5 Energy Thresholds and Recursive Rules

A comparison among the threshold energies for the deformed metrics $g_{D,\text{int}}(E)$ of the electromagnetic, gravitational, leptonic, hadronic interactions, given by Eqs. (2.3), (2.7), (2.15), (2.20), yields to an increasing order of $E_{0,\text{int}}$ from the electromagnetic to the hadronic interaction:

$$E_{0\text{e.m.}} < E_{0\text{grav}} < E_{0\text{lep}} < E_{0\text{had}}. \tag{2.21}$$

Moreover, it can be easily seen that a proportion exists among them provided that they are coupled in order of growing energy:

$$\frac{E_{0\text{grav}}}{E_{0el}} = 4.49 \pm 0.02; \qquad \frac{E_{0,\text{had}}}{E_{0,\text{lep}}} = 4.57 \pm 0.01, \tag{2.22}$$

namely,

$$\frac{E_{0\text{grav}}}{E_{0el}} \simeq \frac{E_{0,\text{had}}}{E_{0,\text{lep}}}. \tag{2.23}$$

These two facts allows us to call this order a 'natural order'. Now, we regroup the four deformed metrics $g_{D,\text{int}}(E)$ in couples following this natural order as sketched in Fig. 2.1. Each couple is composed of the sub-Minkowskian and the over-Minkowskian having the energy thresholds in the same scale of energies. The next couple follows according to the same pattern, regardless if at lower or higher scale of energies for the thresholds, as it can be seen in Fig. 2.1.

A further remark concerns the possible pattern of interactions ensuing from this natural ordering. According to the results summarized in Fig. 2.1, we have two couples of interactions: (I) electromagnetic and gravitational; (II) leptonic and hadronic, ordered by the increasing arrangement of the threshold energies. Moreover, in each couple the former interaction is sub-Minkowskian, and the latter is

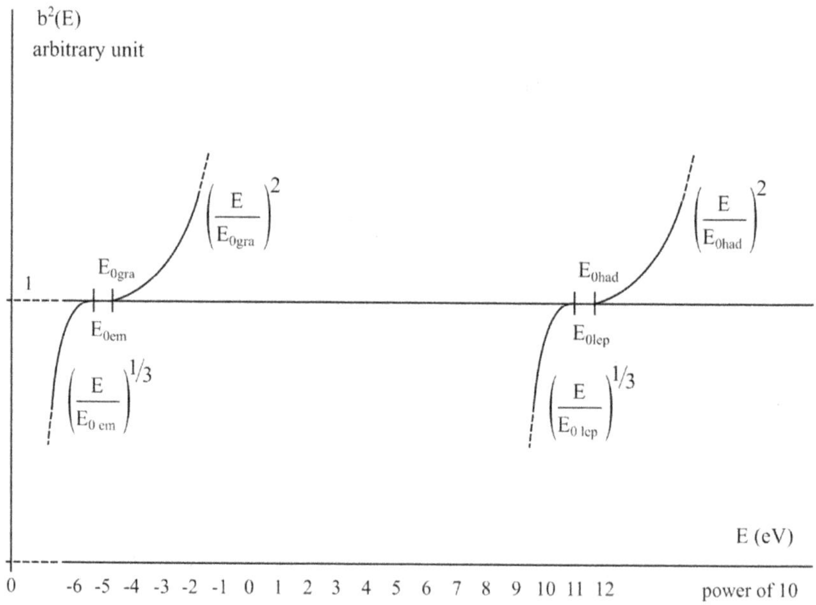

Fig. 2.1. The general pattern of deformed metrics for the four metrics $g_{D,\text{int}}(E)$ coupled according to the natural order.

over-Minkowskian. The fundamental question is whether this pattern ends or not as energy increases. If a third pair exists at higher energies, we can assume that the threshold energies of the new pair, E_{05} and E_{06}, are related to the threshold energies of the previous sub-Minkowskian and over-Minkowskian metrics according to

$$\frac{E_{0,n+2}}{E_{0,n}} = \frac{E_{0,n+4}}{E_{0,n+2}}, \qquad n = 1, 2, \tag{2.24}$$

with $E_{0_{el}} = E_{01}$; $E_{0\text{grav}} = E_{02}$; $E_{0\text{lep}} = E_{03}$; $E_{0\text{had}} = \mathrm{E}_{04}$. In such hypothesis, with the values of the threshold energies for the known interactions, one gets

$$\begin{aligned} E_{05} &\approx 1.3 \times 10^{18}\,\text{GeV}, \\ E_{06} &\approx 6.7 \times 10^{18}\,\text{GeV}. \end{aligned} \tag{2.25}$$

These energy values are quite out of range for the measurement instruments and detectors almost at present, XXI century. Nonetheless this evidence, their effects could affect phenomena measurable in the scale of measure at disposal in the present. By the way one can note that this scale of energies is that one for leptoquark transitions predicted by grand unified and super symmetric grandunified theories, but we do not want to push further this analogy.

Let us also stress that, in the sub-Minkowskian case, the deformed metrics could vanish, in general for an energy value $W_0 \neq 0$, by definition:

$$b_\mu^2(W_0) = 0. \tag{2.26}$$

The physical meaning of the metric vanishing is that, at the energy W_0, the interaction is unable to measure the space time separation. In particular, such a value of the energy corresponds to a point-like object for the interaction considered. Examples of this fact are provided by the inertial mass of the photon for an electromagnetic metric, or the neutrino masses for a leptonic metric. Indeed, for the e.m. interaction, $W_{0,\text{e.m.}}$ can be consistent with the energy value $E_{\min} = 10^{-26}$ microeV considered to be essentially the upper limit for the photon inertial mass, again we do not want to push further this remark.

We now resume all these arguments in the following relations which we call 'recursive rules'. For metrics not homogeneous adjacent and successive we fix l as index for sub-Minkowskian metric and k index for over-Minkowskian metric, therefore the first recursive rule for metrics not homogeneous reads:

$$nE_{0,l} = E_{0,k}, \quad n = 5,$$

where the value of n is taken from the phenomenology.

For metrics homogeneous and successive, we fix only the index i. In the case of sub-Minkowskian metric $i = l$, in the case of over-Minkowskian metric $i = k$, therefore the second recursive rule for homogeneous metrics reads:

$$rE_{0,i} = E_{0,i+1}, \quad r = 2 \cdot 10^{16},$$

where the value of r is taken from the phenomenology.

A representation of the recursive pattern for the deformed metrics according to the recursive rules is illustrated in Fig. 2.2, where due to the lack of any limit at low as well at high energies they are intended as asymptotic. A possible stop to this recursive pattern could be given by two simple concepts: the zero energy and the maximum energy attainable in the Universe. Even if the zero energy is a concept well acceptable, despite the true difficulty in measuring a zero energy, the concept of the maximum energy in the Universe is quite impossible

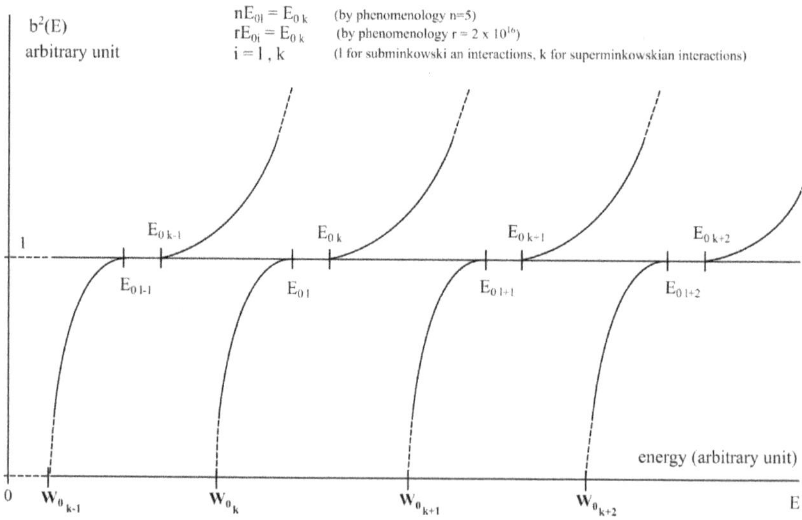

Fig. 2.2. Qualitative behaviour of recursive asymptotic metrics.

to make independent from conditioning hypotheses. But we could use the 'Kostro constant' to have an idea of the upper limit to the recursive metrics. This constant, named after L. Kostro, who calculated it [8], gives the estimate of the 'maximum energy attainable' in the present Universe at least within the Newtonian mechanics as limit of the Schwarzschild metric in general relativistic sense. Anyway we can consider these two recursive rules only for their predictive power, it means that they are useful and easy to handle tools to looking for interactions other than the known ones, and perhaps the one and only way to find new interactions existing in geometrical sense.

References

[1] A. Enders, G. Nimtz. *J. Phys. I (France), 3*, 1089, 1993.

[2] C. O. Alley. *Quantum Optics, Experimental Gravity and Measurement Theory*, P. Meystre, M. Scully Eds. Plenum Press, New York, 1983, p. 363.

[3] P. S. M. de La Place. *Mécanique Céleste, 5th volume* translated with commentary by N. Bodwitch, Boston 1829, the evaluation of the speed of the gravitational action is obtained in the system Sun Earth Moon by the lunar liberation motion.

[4] W. D. Walker, J. Dual *Gravitational Waves Proceeding of 2nd E. Amaldi Conference*, E. Coccia, G. Pizzella, G. Veneziano Eds. World Scientific, Singapore, 1997.

[5] T. van Flandern. *Phys. Lett. A*, 250, 1, 1998.

[6] T. van Flandern, J. P. Vigier. *Found. Phys.*, 32, 1031, 2002.

[7] F. Cardone *et al. New Adv. Phys.*, 3, 1, 7, 2009.

[8] L. Kostro, B. Lange. *Phys. Essays* 10, 1999.

Part II

Five Dimensional Metrics

Chapter 3

Einstein Equations

3.1 Five Dimensional Einstein Equations and Solutions

The vacuum Einstein equations in the five dimensional Riemann space R_5 are

$$R_{AB} - \frac{1}{2} g_{AB}^{(5)} R = \Lambda g_{AB}^{(5)}, \tag{3.1}$$

where R_{AB} and $R = R_A^A$ are the five dimensional Ricci tensor and scalar curvature, respectively, while Λ is the so called cosmological constant. The five dimensional deformed metric is defined by:

$$g_D^{(5)}(E) = \mathrm{diag}(b_0^2(E), \quad b_1^2(E), \quad b_2^2(E), \quad b_3^2(E), \quad b_5^2(E)).$$

In general $b_5^2(E)$ is the metric element corresponding to the fifth coordinate the energy E, but for dimensional reasons we consider the coordinate vector as

$$(ut, x, y.z, l_0 E) \quad \text{or rather} \quad (x_0, x_1, x_2, x_3, x_5),$$

where u is the maximal causal speed of the interaction described by the metric $g_D^{(5)}(E)$ and l_0 is the 'Pessa constant', so named after Eliano Pessa who introduced it, which assures the dimensional conversion of energy to length. At a first glance l_0 has the dimensions of the inverse of a linear energy density. The concept of energy density will be of groving relevance in the next development of Deformed Space Time theory.

The factor Λ is assumed to be a genuine constant, although in principle it might also depend on both the energy E and the space time coordinates $x : \Lambda(x, E)$.

In the case of spatial isotropy, where we set $b_1(E) = b_2(E) = b_3(E) = b(E)$, these equations can be reduced to the following form [1]–[3]:

$$\begin{cases} 3(-2b''b_5 + b'b_5') = 4\Lambda b{b_5}^2, \\ b_5[b_0^2(b')^2 - 2b_0b_0'bb' - 4b_0^2bb'' - 2b_0b_0''b^2 + b^2(b_0')^2] \\ \quad + b_0bb_5'(2b_0b' + b_0'b) = 4\Lambda b_0^2b^2b_5^2, \\ 3b'(b_0b)' = -4\Lambda b_0b^2b_5, \end{cases} \tag{3.2}$$

where a prime denotes derivation with respect to E and we put $c = l_0 = 1$.

Since Λ is related to the vacuum energy in General Relativity and experimental evidence shows that $\Lambda \simeq 3.10^{-52}\mathrm{m}^{-2}$, it is possible to assume $\Lambda = 0$ as far as one is not interested in quantum effects. Many are the solutions corresponding to $\Lambda = 0$, among them there is:

$$\begin{cases} b_5(E) = const., \\ b_0(E) = \left(1 + \frac{E}{E_0}\right)^2, \\ b(E) = const., \end{cases} \tag{3.3}$$

where E_0 is a constant having the dimensions of energy. This solution is in fact the unique solution corresponding to a constant metric element $b_5(E)$ in a larger family in which $b_0(\mathrm{E})$ is arbitrary and the couple of metric elements (b_5, b_0) satisfy the ordinary differential equation:

$$-2b_0b_0''b_5 + (b_0')^2b_5 + b_0b_0'b_5' = 0. \tag{3.4}$$

Moreover, the metric (3.3) coincides with the phenomenological gravitational metric in the hypothesis of spatial isotropy.

Other interesting classes of solutions to Eqs. (3.2) for $\Lambda = 0$ can be obtained by the ansatz that all the metric elements in $g_{\mu\nu}(\mathrm{E})$ are pure powers of E:

$$\begin{cases} b_0(E) = (E/E_0)^q, \\ b_1(E) = (E/E_0)^m, \\ b_2(E) = (E/E_0)^n, \\ b_3(E) = (E/E_0)^p, \end{cases} \tag{3.5}$$

while for the metric element $b_5(E)$ one assumes coherently,

$$b_5(E) = E^r. \tag{3.6}$$

In this case, the Einstein equations reduce to the algebraic system:

$$\begin{cases} (2+r)(p+m+n) - m^2 - n^2 - p^2 - mn - mp - np = 4\Lambda E^{r+2}, \\ (2+r)(p+q+n) - n^2 - p^2 - q^2 - np - nq - pq = 4\Lambda E^{r+2}, \\ (2+r)(p+q+m) - m^2 - p^2 - q^2 - mp - mq - nq = 4\Lambda E^{r+2}, \\ (2+r)(p+m+n) - m^2 - n^2 - q^2 - mn - mq - nq = 4\Lambda E^{r+2}. \\ mn + mp + mq + np + nq + pq = -4\Lambda E^{r+2}, \end{cases} \tag{3.7}$$

which admits at least twelve classes of solutions. They include as special cases **all** the phenomenological metrics discussed in Part I. In particular, the metric (3.3) can be obtained as the only metric which lives in the intersection of three of the relevant classes, namely it is obtained by setting:

$$q = 2, \quad m = n = p = r = 0, \tag{3.8}$$

which obviously reduces to gravitational metric $g_{\text{grav.}}(E)$ by a rescaling and a translation of the energy parameter E_0.

The fact that it is possible, by the mere consideration of the Einstein equations in vacuum, to recover all the phenomenological metrics for the four fundamental interactions, is a further confirmation of the fact that deformed space time in four dimensions is indeed naturally embedded in the five dimensional scheme of the space time energy. Now the problem moves to the quest for the fifth metric element. Even with the assumption of the Eq. (3.6) the search for the exponent r remain an open question, for the moment. The first attempt to solve this quest was by the use of the isometries in the energy metrics, since at a first glance seemed the natural way to perform this research and the natural mathematical tool was the Killing Equations.

3.2 Discussion on Five Dimensional Killing Equations and Their Solutions

Now we introduce the Killing symmetries of the space R_5 [4].

In R_5, the Lie derivative L of a rank-2 covariant tensor field T_{AB} along the five-vector field $\tilde{\xi} = \{\xi_A(x, x^5) \equiv \xi_A(x^B)\}$ is given as usual by

$$L_{\tilde{\xi}} T_{AB} = T_A^C \xi_{C;B} + T_B^C \xi_{C;A} + T_{AB;C} \xi^C \tag{3.9}$$

with '$;A$' denoting covariant derivative with respect to x^5. If the tensor coincides with the metric tensor g_{AB}, whose covariant derivative vanishes, its Lie derivative becomes

$$L_{\tilde{\xi}} g_{AB} = \xi_{A;B} + \xi_{B;A} = \xi_{[A;B]}, \tag{3.10}$$

where the bracket $[\cdot\cdot]$ means symmetrization with respect to the enclosed indices.

Then, a five-vector $\tilde{\xi}$ is a Killing vector if the Lie derivative of the metric tensor with respect to ξ vanishes

$$L_{\tilde{\xi}} g_{AB} = 0 \Leftrightarrow \xi_{[A;B]} = 0 \Leftrightarrow \xi_{A;B} + \xi_{B;A} = 0 \tag{3.11}$$

are the Killing equations in R_5. Since the Lie derivative is nothing but the generalization of directional derivative, this means that the Killing vectors correspond to isometric directions. The integrability conditions of (3.10) are given by:

$$\xi_{A;BC} = \xi_{C;[BA]} = R^D_{CBA} \xi_D \Leftrightarrow L_{\tilde{\xi}} \Gamma^A_{BC} = 0, \tag{3.12}$$

where R_{ABCD} and Γ^A_{BC} are the five dimensions Riemann Christoffel tensor and affine connection respectively. The Eq. (3.12) are integrable under the conditions

$$L_{\tilde{\xi}} R_{ABCD} = 0. \tag{3.13}$$

For metric $g_D^{(5)}(E)$, from the Christoffel symbols Γ^A_{BC} of the metric $g_{AB,D5}(x^5)$, (3.11) take the form of the following system of 15 coupled, partial derivative differential equations in R_5 for the Killing vector $\xi_A(x^B)$:

$$b_5(x^5)\xi_{0,0}(x^A) \pm b_0(x^5)b_0'(x^5)\xi_5(x^A) = 0, \tag{3.14}$$

$$\left.\begin{aligned} \xi_{0,1}(x^A) + \xi_{1,0}(x^A) = 0 \\ \xi_{0,2}(x^A) + \xi_{2,0}(x^A) = 0 \\ \xi_{0,3}(x^A) + \xi_{3,0}(x^A) = 0 \end{aligned}\right\} \text{type I conditions}, \tag{3.15}$$

$$b_0(x^5)(\xi_{0,5}(x^A) + \xi_{5,0}(x^A)) - 2b_0'(x^5)\xi_0(x^A) = 0\} \quad \text{type II condition}, \tag{3.16}$$

$$b_5(x^5)\xi_{1,1}(x^A) \mp b_1(x^5)b_1'(x^5)\xi_5(x^A) = 0, \tag{3.17}$$

$$\left.\begin{aligned} \xi_{1,2}(x^A) + \xi_{2,1}(x^A) = 0 \\ \xi_{1,3}(x^A) + \xi_{3,1}(x^A) = 0 \end{aligned}\right\} \text{type I conditions}, \tag{3.18}$$

$$b_1(x^5)(\xi_{1,5}(x^A) + \xi_{5,1}(x^A)) - 2b_1'(x^5)\xi_1(x^A) = 0\} \quad \text{type II condition}, \tag{3.19}$$

$$b_5(x^5)\xi_{2,2}(x^A) \mp b_2(x^5)b_2'(x^5)\xi_5(x^A) = 0, \tag{3.20}$$

$$\xi_{2,3}(x^A) + \xi_{3,2}(x^A) = 0\} \text{ type I conditions}, \tag{3.21}$$

$$b_2(x^5)(\xi_{2,5}(x^A) + \xi_{5,2}(x^A)) - 2b_2'(x^5)\xi_2(x^A) = 0\} \quad \text{type II condition}, \tag{3.22}$$

$$b_5(x^5)\xi_{3,3}(x^A) \mp b_3(x^5)b_3'(x^5)\xi_5(x^A) = 0, \tag{3.23}$$

$$b_3(x^5)(\xi_{3,5}(x^A) + \xi_{5,3}(x^A)) - 2b_3'(x^5)\xi_3(x^A) = 0\} \quad \text{type II condition}, \tag{3.24}$$

$$2f(x^5)\xi_{5,5}(x^A) - f'(x^5)\xi_5(x^A) = 0, \tag{3.25}$$

where now, 'A' denotes ordinary derivative with respect to x^A.

Equations (3.24)–(3.25) can be divided in 'fundamental' equations and 'constraint' equations (of type I and II). The earlier system is in general overdetermined, i.e. its solutions will contain numerical coefficients satisfying a given algebraic system. Explicitly solving it yields:

$$\xi_\mu(x^A) = F_\mu(x^{A\neq\mu}) \pm (-\delta_{\mu 0} + \delta_{\mu 1} + \delta_{\mu 2} + \delta_{\mu 3}) b'_\mu(x^5)(b_5(x^5))^{1/2} \times \int dx^\mu F_5(x), \quad (3.26)$$

$$\xi_5(x^A) = (b_5(x^5))^{1/2} F_5(x). \quad (3.27)$$

The five unknown functions $F_A(x^{B\neq A})$ are restricted by the two following types of conditions:

(I) Type I (Cardinality 4, $\mu \neq \nu \neq \rho \neq \sigma$):

$$\pm A_\mu(x^5)G_{,\nu\rho\sigma}(x) + B_\mu(x^5)G_{,\mu\mu\nu\rho\sigma}(x) + b_\mu(x^5)F_{\mu,5}(x^{A\neq\mu}) - 2b'_\mu(x^5)F_\mu(x^{A\neq\mu}) = 0. \quad (3.28)$$

(II) Type II (Cardinality 6, symm. in $\mu, \nu, \mu \neq \nu \neq \rho \neq \sigma$):

$$F_{\mu,\nu}(x^{A\neq\mu}) + F_{\nu,\mu}(x^{A\neq\nu}) + \pm(-\delta_{\mu 0} + \delta_{\mu 1} + \delta_{\mu 2} + \delta_{\mu 3}) b_\mu(x^5) b'_\mu(x^5)(b_6(x^5))^{-1/2} G_{,\nu\nu\rho\sigma}(x) +, \pm(-\delta_{\nu 0} + \delta_{\nu 1} + \delta_{\nu 2} + \delta_{\nu 3}) b_\nu(x^5) b'_\nu(x^5)(b_5(x^5))^{-1/2} G_{,\mu\mu\rho\sigma}(x) = 0, \quad (3.29)$$

where we introduced the fake four-vectors excluding the Einstein summation convention on repeated indexes. Indeed it is in general (i=1, 2, 3) that the Einstein summation convention on repeated indexes holds

$$b_\mu(x^5) = g_{\mu\mu,D5}(x^5)(\delta_{\mu 0} - \delta_{\mu i})$$
$$b_{5_\mu}(x^5) = \pm g_{44,D5}(x^5)$$

which clearly show the nonvector nature of $A_\mu(\mathrm{x}^5)$ and $B_\mu(\mathrm{x}^5)$ in the four dimensional subspaces of R_5,

$$A_\mu(x^5) = (-\delta_{\mu 0} + \delta_{\mu 1} + \delta_{\mu 2} + \delta_{\mu 3}) b_\mu(x^5)(b_5(x^5))^{-1/2} \times \left[-(b'_\mu(x^5))^2 + b_\mu(x^5) b''_\mu(x^5) - \frac{1}{2} b_\mu(x^5) b'_\mu(x^5) b'_5(x^5)(b_5(x^5))^{-1} \right], \quad (3.30)$$

$$B_\mu(x^5) = b_\mu(x^5)(b_5(x^5))^{1/2}, \quad (3.31)$$

and defined the function

$$G(x) = \int d^4 x F_5(x). \tag{3.32}$$

So, the problem is reduced to finding the function F_5 in order to get the fifth metric element $b_5^2(E)$.

For five dimensional sub-Minkowskian metrics electromagnetic and leptonic, we obtain:

$$g_{AB,D5}(x^5) = \text{diag}\left(1, -\left(\frac{x^5}{x_0^5}\right)^{1/3}, -\left(\frac{x^5}{x_0^5}\right)^{1/3} - \left(\frac{x^5}{x_0^5}\right)^{1/3}, \right.$$
$$\left. \pm\left(6c\left(\frac{x^5}{x_0^5}\right)^{5/3} + \gamma\left(\frac{x^5}{x_0^5}\right)^2\right)^{-1}\right), \tag{3.33}$$

where x_0^5 means the energy threshold for electromagnetic or leptonic interaction.

For five dimensional over-Minkowskian metric hadronic we obtain:

$$g_{AB,D5}(x^5) = \text{diag}\left(\left(\frac{x^5}{x_0^5}\right)^2, -\frac{2}{25}, -\frac{4}{25}, -\left(\frac{x^5}{x_0^5}\right)^2, \pm\frac{1}{\gamma\left(\frac{x^5}{x_0^5}\right)^2 + C}\right), \tag{3.34}$$

where x_0^5 means the energy threshold for hadronic interaction, and the constants C, γ are real numbers.

For five dimensional over-Minkowskian metric gravitational we obtain:

$$b_5(x^5) = \frac{1}{\gamma(x^5 + x_0^5)^2 + C}, \quad C, \gamma \in R, \tag{3.35}$$

where x_0^5 means the energy threshold for gravitational interaction, and the constants C, γ are real numbers.

As we shall see the Eqs. (3.33) and (3.35) give expression for the fifth metric element $b_5^2(E)$ neither physically realistic nor phenomenological verified for the metric of any interaction.

We report again here below in the following paragraph the $b_5(E)$ for electromagnetic, leptonic, hadronic and gravitational metrics in order to stress their deep difference with respect to those ones we have obtained in the next Chapter 4 by the Ricci flow method.

3.3 Quest for the Fifth Metric Element

The quest for the fifth element in the metrics for the fundamental interactions has been performed using the isometries of the metrics according the Killing equations [4] as shown in Section 3.2. This method is truly complicate and cumbersome. The results for the solution of the Killing equations have been obtained within the hypothesis that the elements of the metrics are functions given by a power of the energy. Moreover to solve the Killing equations it was necessary to use extra mathematical hypothesis giving rise to the conditions of solvable equations. The results are summarized in formulae which did not find any match with the physical phenomenology and did not give any useful contribution to study the new phenomena connected to the deformed space time, as for example the nuclear metamorphosis and its consequences. Anyway for the sake of completeness we quote such results which had no experimental match. We report here below these fifth elements of the metrics coming from Killing equations using the same notation as in Section 3.2: for electromagnetic and leptonic metric

$$g_{AB,D5}(x^5) = \mathrm{diag}\left(1, -\left(\frac{x^5}{x_0^5}\right)^{1/3}, -\left(\frac{x^5}{x_0^5}\right)^{1/3}, -\left(\frac{x^5}{x_0^5}\right)^{1/3}, \right.$$
$$\left. \pm\left(6c\left(\frac{x^5}{x_0^5}\right)^{5/3} + \gamma\left(\frac{x^5}{x_0^5}\right)^{2}\right)^{-1}\right), \tag{3.36}$$

where the constant belonging to real positive numbers; for hadronic metric

$$g_{AB,D5}(x^5) = \mathrm{diag}\left(\left(\frac{x^5}{x_0^5}\right)^2, -\frac{2}{25}, -\frac{4}{25}, -\left(\frac{x^5}{x_0^5}\right)^2, \pm\frac{1}{\gamma\left(\frac{x^5}{x_0^5}\right)^2 + C}\right), \tag{3.37}$$

where again the constant belonging to real positive numbers;

for gravitational metric

$$(x^5) = \frac{1}{\gamma(x^5 + x_0^5)^2 + C}, \quad C, \gamma \in R. \tag{3.38}$$

For the sake of clearness, we remind that x_0^5 stands for the energy threshold for the interaction treated.

The conclusion was that it was necessary to look for other ways more grounded on physical motivations.

We therefore acquired the result that considering isometries of energy metrics is not the right way in the quest for the fifth metric element. Then we changed the concept of isometries with the volume conservation since at the very basis of matter is the constant density in the nuclei. We considered that any deformation producing a non flat nuclear space time could change the measure and the shape of the volume occupied by the nuclei themselves, but the volume preserving its value, like in an affine transform following the Cavalieri theorem. So we looked for a mathematical tool preserving the volume also in a space time deformed and identified it in the method of the Ricci Flow. Besides considerations on matter, there is the need to have a preserved volume also when considering energy density which gives rise to space time deformation ruling new phenomena as we shall see. In fact it is possible to state a theorem which assures that the Ricci flow preserves the volume. We report the enunciation of this theorem which can be found say in [5]:

- 'if in a metric flow remains constant with respect to the evolution parameter t the mean value of a symmetric tensor S_{ij} function of coordinates and evolution parameter t, then remains also constant the volume V_D',
- the volume V_D is defined in a n dimensional manifold M_n where exists a coordinate domain D contained in M_n which is a deformed hyperparallelepipedon whose edges are segments of coordinate lines.

The volume V_D is the volume inside the deformed space time. Obviously ever in the framework of the energy metrics the evolution parameter is nothing else than the coordinate x_5 or rather the energy E assumed as variable, too.

In this sense the energy E assumes three meanings: phenomenological variable, coordinate, evolution parameter.

References

[1] F. Cardone *et al. Gen. Rel. Grav.* 30, 1619, 1998.
[2] M. Francaviglia *et al. Gen. Rel. Grav.* 31, 1049, 1999.

[3] M. Francaviglia *et al. Found. Phys. Lett.* 12, 281, 347, 1999.

[4] A. Marrani. ‘*Aspetti matematici della Relatività Deformata in cinque dimensioni’* (in Italian) Ph.D. Thesis in Physics, R. Mignani. Supervisor, University Rome Tre, Rome 2003.

[5] S. Benenti and F. Cardone. *Ann. Math. Phys.*, 7, 1, 024–053, 2024.

Chapter 4

Ricci Flows of Five Dimensional Energy Metrics

4.1 Volume Conservation and Ricci Flow

We set the following definitions we are going to use. Given an n dimensional manifold M_n with coordinates $(x) = (x_1, x_2, \ldots, x_n)$, on this manifold we consider a coordinated domain $D \subset M_n$, which is a deformed hyperparallelepipedon whose edges are segments of coordinate lines. To such a domain we apply the derivation theorem under the integral sign: in fact, the results obtained below are also valid in more general cases. We call 'metric flow' a family of metric tensors $g_{ij}(x,t)$ defined over D, depending on an evolution parameter t such that it satisfies flow equations of the type

$$\partial_t g_{ij}(x,t) = -S_{ij}(x,t) + \frac{1}{n}\overline{S}(t) g_{ij}(x,t), \tag{4.1}$$

where $S_{ij}(x,t)$ is a symmetric tensor defined on D and

$$\overline{S} = \frac{1}{V_D}\int_D S dV = \frac{1}{V_{\mathrm{D}}}\int_D S\sqrt{|g|}dx \tag{4.2}$$

is the mean value over D of the scalar S given by the standard scalar product $g^{ij}S_{ij}$.

Now it is possible to state a theorem which assures that the Ricci flow preserves the volume. We report the enunciation of this theorem without proof.

4.1.1 *Theorem of volume conservation*

'If in a metric flow (4.1) the mean value $\overline{S}$ remains constant with respect to the evolution parameter t then the volume V_D remains constant'.

We will discuss the physical meaning and relevance of this theorem in the next.

It is also possible to give a definition of normalized Ricci flow: it is a family of metrics $g_{ij}(x,t)$ parametrized by an independent evolution variable t and defined over a domain D of a Riemannian manifold M_n and satisfying the normalized flow equations

$$\partial_t g_{ij}(x,t) = -R_{ij}(x,t) + \frac{1}{n}\overline{R}(t) g_{ij}(x,t), \tag{4.1}$$

where $R_{ij}(x,t)$ is the Ricci tensor of the metric $g_{ij}(x,t)$ and $\overline{R}$ is the mean value of the Ricci scalar R in the domain D:

$$\overline{R} = \frac{1}{V_D}\int_D R dV = \frac{1}{V_D}\int_D R\sqrt{|g|}dx \tag{4.3}$$

$$R = g^{ij} R_{ij}. \tag{4.4}$$

It is possible to calculate that all energy metrics of the interactions have the integral of Eq. (4.4) equal to zero, therefore to postulate the existence of a normalized Ricci flow for any energy metrics necessarily implies $\overline{R} = 0$. In other words from the hypothesis that the energy metrics admit a normalized Ricci flow it necessarily follows $\overline{R} = 0$, which means that the Ricci flow of any energy metric of the fundamental interactions is in fact not normalized.

With $\overline{R} = 0$ the equations of the normalized flow (4.3) reduce to those of a non-normalized Ricci flow

$$\partial_t g_{ij} = -R_{ij}. \tag{4.5}$$

It is well known that the existence of a normalized Ricci flow is a sufficient condition for the conservation of the volume of the definition domain D, but it is not necessarily a 'necessary condition'.

However, we observe that the conservation of the volume V_D is still guaranteed by the 'Theorem of Volume Conservation' according to which if in a flow of metrics defined by Eq. (4.1) the mean value $\overline{S}$ does not vary when t varies then the volume V_D also remains constant. Since $\overline{S}$ correspond to $\overline{R}$ and we have for energy metrics that $\overline{R} = 0$, we conclude that this condition is satisfied.

The volume V_D is the volume inside the deformed space time, and obviously ever in the framework of the energy metrics the evolution parameter t is nothing else than the coordinate x_5 or rather the energy E. We now are assured that the way chosen, the Ricci flow, satisfies in any case the volume conservation in the deformed space time as the fifth coordinate, the energy, changes by. Therefore we also are in the condition to be confident that the nuclear density is constant even in the deformed space time. This fact allows us to go on by the method of the Ricci flow to calculate the fifth metric element for all the energy metrics of the fundamental interactions.

4.2 The Fifth Metric Element for Energy Metrics

We report the results of the calculation of the fifth metric element for the energy metrics of each interaction. These results allow to get the true five dimensions energy metrics for interactions.

We remember that in the coordinates $(x_0, x_1, x_2, x_3, x_5)$ the energy metric tensor and the Ricci tensor are both diagonalized, this fact simplifies the differential equations whose solutions give us the fifth elements of energy metrics, which are listed below in the following order, first the over-Minkowskian interactions hadronic and gravitational then the sub-Minkowskian interaction leptonic and electromagnetic, regardless the order set by the recursive rules.

The hadronic interaction metric above the threshold $E_{0,\text{had}}$ admits a non normalized Ricci flow as long as the function $b^2_{5,\text{had}}(E)$ satisfies the differential equation

$$x_5(b^2_{5,\text{had}})' - 6b^2_{5,\text{had}} = 0, \tag{4.6}$$

where the derivation is respect to the energy E corresponding to the x_5 and whose complete integral is

$$b^2_{5,\text{had}} = K_{\text{had}} x_5^6, \quad K_{\text{had}} > 0 \textit{ constant}, \tag{4.7}$$

where K_{had} is an arbitrary positive constant.

The gravitational interaction metric above the threshold $E_{0,\text{grav}}$ admits a non normalized Ricci flow as long as the function $b^2_{5,\text{grav}}(E)$

satisfies the differential equation

$$(E_{0\text{grav}} + x_5)(b^2_{5,\text{grav}})' - 6b^2_{5,\text{grav}} = 0, \tag{4.8}$$

where the derivation is respect to the energy E corresponding to the x_5 and whose complete integral is

$$b^2_{5,\text{grav}} = K_{\text{grav}}(E_{0,\text{grav}} + x_5)^6, \quad K_{\text{grav}} > 0 \text{ } \textit{constant}, \tag{4.9}$$

where K_{grav} is an arbitrary positive constant.

The leptonic interaction metric below the threshold $E_{0,\text{lep}}$ admits a non normalized Ricci flux as long as $b^2_{5,\text{lep}}$ is constant:

$$b^2_{5,\text{lep}} = k_{\text{lep}}, \quad k_{\text{lep}} > 0 \text{ } \textit{dimensionless constant.} \tag{4.10}$$

The electromagnetic interaction metric below the threshold $E_{0,\text{e.m.}}$ admits a non-normalized Ricci flux as long as $b^2_{5,\text{e.m.}}$ is constant:

$$b^2_{5,\text{e.m.}} = k_{\text{e.m.}}, \quad k_{\text{e.m}} > 0 \text{ } \textit{dimensionless constant.}$$

We stress that the evidence of the similar behaviour of the two fifth elements for leptonic and electromagnetic metrics means that these two sub-Minkowskian metrics have similar energy gauge, or rather 'measure' the energy in the 'similar' way regardless the energy value, but not in 'identical' way unless we find reasons to set $k_{\text{e.m.}} = k_{\text{lep}}$. We remark that this occurrence gives us another reason to understand the so called 'mixing' of these two interactions, besides the fact that for energies $E \geq E_{0,\text{lep}}$ both metrics assume the form of flat Minkowski metric. So now we can say that for energies above the leptonic threshold the electromagnetic and leptonic interactions measure the space time intervals in the same Minkowskian way and if $k_{\text{e.m.}} = k_{\text{lep}}$ i.e. $b^2_{5,\text{e.m.}} = b^2_{5,\text{lep}}$ they measure also the energies in the same way.

4.3 Pessa Constant and Convention

We used for the energy the fifth coordinate x_5 and introduced the Pessa constant l_0 in order to get the right spatial dimension when calculating say the space time energy interval ds^2. Now we explore other ways to use the Pessa constant l_0. We consider the dimensionless variable given by the ratio

$$\frac{x_5}{E_{0\text{int}}} \tag{4.11}$$

between the energy coordinate x_5 and the energy threshold $E_{0,\text{int}}$ characteristic for each interaction. It is useful to recall that the symbol *int* (interaction) stands obviously for: *e.m., grav., lep., had.*

We redefine the variable x_5, which is our energy coordinate, introducing a new Pessa constant L, which has the dimensions of a length, preserving the dimensions of a length for x_5. At the same time we introduce the new variable

$$\overline{x}_5 = L \cdot \frac{x_5}{E_{0\text{int}}}, \tag{4.12}$$

so that the dimensionless variable is expressed by

$$\frac{x_5}{E_{0\text{int}}} = \frac{\overline{x}_5}{L}. \tag{4.13}$$

Then we set the following definition as Pessa convention:

$$k \equiv \frac{L}{E_{0\text{int}}}, \tag{4.14}$$

where k has the dimensions of inverse linear energy density as was the old Pessa constant l_0. We merely shown here another way for making the fifth energy coordinate x_5 dimensionally homogeneous to the others through the introduction of the new Pessa constant L, whose meaning, physical identification and numerical value are reported below.

The value of L is in the range 4–8 μm as it was originally calculated theoretically and reported in [1]. This L is the characteristic linear dimension for all interactions and whose volume L^3 allows us to calculate the critical energy density $D_{C,\text{int}}$ which gives rise to the metamorphosis of matter. In general for each interaction this critical energy density is given by:

$$D_{C,\text{int}} = \frac{E_{0,\text{int}}}{L^3}. \tag{4.15}$$

From the phenomenology of experiments concerning both nuclear particle emissions in deformed space time conditions and transformations as nuclear metamorphoses, we can use $L = 10\,\mu$m which corresponds to the characteristic diameter measured for the so-called Ridolfi cavities [2]. We return to this problem of the critical energy density $D_{C,\text{int}}$ in the next.

4.4 Remarks on Metrics

It is worth highlighting once more the analogy between the hadronic and gravitational metrics. In both cases a deformation of the temporal coordinate occurs. Furthermore, one of the spatial parameters, which we have conventionally assumed as the third parameter, varies with energy like the temporal one in an over-Minkowskian way, that is, it approaches the Minkowskian limit for energy values greater than the energy threshold of the interaction. In the hadronic case the other two spatial parameters are constant, but of different value. Then the three dimensional space is anisotropic for the hadronic interaction, even in its flat part for the other metric elements, see its behaviour inside the atomic nucleus due the existence of non-spherical nuclei. Even more the fifth element of both these metrics, corresponding to energy coordinate, shows the same dependence from the sixth power of energy, see Eqs. (4.7) and (4.9), that means any displacement say of $1/10$ from the energy threshold can change of a factor 10^6 the behaviour of the interaction and its consequences at phenomenological level. As we will see this fact has been tested for the hadronic interaction in the phenomenon of nuclear metamorphosis. This also means that both hadronic and gravitational interactions have the same 'energy gauge'.

We remind once more that the threshold energy, generally indicated by $E_{0,\text{int}}$ is the energy value at which the metric parameters of the interactions reach a constant value and the metric becomes flat Minkowskian.

Note that for both electromagnetic and leptonic interactions the metric is isochronous, the time parameter does not change as the energy varies, furthermore it is spatially isotropic and sub–Minkowskian, it approaches the Minkowski limit for values increasing in energy but less than the threshold energy. Their fifth metric element of both these metrics, corresponding to energy coordinate, shows the same behaviour as constant $k_{\text{e.m.}}$, k_{lep}, this also means that both e.m. and leptonic interactions have the same 'energy gauge'.

We consider the possibility that $k_{\text{e.m.}} = k_{\text{lep}} = 1$ in this case they are also 'isoenergetic' or rather they measure the energy in identical way.

4.5 Physical and Phenomenological Identification of the Conserved Volume

The theorem of the volume conservation in Section 4.1 has a general character, it is independent from the normalized Ricci flow, it does not depend on the metrics of the interactions but is applied to them to verify that the interaction conserves the volume. In particular, its application to the hadronic metric allows us to conclude that it conserves volume. Likewise, the interaction represented by this metric conserves the volume, therefore one of the main characteristics of the nuclear interaction, hadronic interaction in the nucleus, which is the constant of the density in the nucleus, is respected even if the nucleus is subjected to 'inner' deformation beyond those already known for ellipsoidal nuclei. In fact, the conservation of the volume, regardless of the deformation, allows the nuclear density to be constant.

In order to identify the conserved volume we evaluate the physical volume V_D of the deformed hyperparallelepiped. In general we estimate V_D at the energy thresholds E_0 counting on making a useful estimate not only for the metric of the hadronic interaction but also for the other interactions and related metrics.

Let us now present our proposals regarding the physical volume of the deformed hyperparallelepiped whose edges are segments of coordinated lines, in the following setting L equal to Pessa constant, as defined in Section 4.3, t is the time corresponding to the coordinate x_0, and $u = u(E)$ the maximal causal speed for the interaction, we remind that $u(E) = u_{\text{int}}(E) = c(b_{0,\text{int}}(E)/b_{k,\text{int}}(E)$ where $k = 1, 2, 3$, and c is the light speed in vacuum.

First case, three dimensions: we evaluate the volume of this hyperparallelepiped through the following identification with the critical volume for nuclear metamorphosis, this identification is valid in a three dimensional Euclidean space with only three spatial coordinates, i.e. for the spatial part of the Minkowskian metric representation of an interaction, so we set $V_D = V_C = L^3$, where V_D is the volume of the domain D, V_C is the critical volume and L is the aforementioned Pessa constant. It is clear that this estimate, based on this identification, can be valid for any interaction metric around the threshold energy $E_{0,\text{int.}}$ which, we remember, is a point of discontinuity in the representation by the Heaviside function.

Second case, four dimensions.

First possibility: we remember that $x_0 = u(E)t, [x_0] = [\text{length}]$, by setting $E \to E_0$, let's set $x_0 \propto c(h/E_0)$, for $E \to E_0, x_0 = u(E)t \to c(h/E_0)$. Therefore we estimate that $V_D = L^3 c(h/E_0)$, $c = u(E_0)$, h the Planck constant, from which $[V_D] = [L^3 c(h/E_0)] = [\text{length}]^4$. This estimate is based on the fact that the Plank's constant can be identified with the constant of the first integral of the geodesic motion referred to the coordinates time and energy [3].

Second possibility: by setting $E \to E_0$, let's set $x_0 \propto (e^2/E_0)$, for $E \to E_0, x_0$ tends to (e^2/E_0), so we estimate that $V_D = L^3(e^2/E_0), e^2$ square of the elementary electric charge, we remember that e is constant and relativistically invariant in Minkowskian space, from which follows $[V_D] = [L^3(e^2/E_0)] = [\text{length}]^4$. This estimate is based on the fact that the square of the elementary electric charge could be identifiable with a constant of the first integral of the geodesic motion referred to the space and energy coordinates [3].

Third case, five dimensions.

First possibility: again we remember that $x_0 = u(E)t, [x_0] = [\text{length}], x_5 = L(E/E_0), [x_5] = [\text{length}]$ by setting $E \to E_0$, let's set $x_0 \propto c(h/E_0), i.e.$ for $E \to E_0$, $x_0 = u(E)t$ tends to c, (h/E_0), furthermore for E which tends to E_0 we have $x_5 = L(E/E_0)$ tends to L, where h is Planck constant, L is Pessa constant and $c = u(E_0)$. Therefore we estimate that $V_D = L^3 c(h/E_0) \cdot L = L^4 c(h/E_0)$. Now physical dimensions are: $[\text{V}_D] = [L^3 L c(h/E_0)] = [L^4 c(h/E_0)] = [\text{length}]^5$.

Also this estimate is based on the fact that Plank constant can be identified with the constant of the first integral of the geodesic motion referred to the time and energy coordinates.

Second possibility: by setting $E \to E_0$, let's set $x_0 \propto (e^2/E_0)$, i.e. for $E \to E_0, x_0$ tends to (e^2/E_0). Furthermore for E which tends to E_0 we have $x_5 = L(E/E_0)$ tends to L, Pessa constant, e^2 squared of the elementary electric charge, again we remember that e is constant and relativistically invariant in the Minkowski space. Therefore we estimate that $V_D = L^4(e^2/E_0)$, from which $[V_\mathrm{D}] = [L^4(e^2/E_0)] =$ [length]^5.

Also this estimate is based on the fact that the square of the elementary electric charge could be identifiable with a constant of the first integral of the geodesic motion referred to the space and energy coordinates [3].

It should be noted that all these three-dimensional, four-dimensional and five-dimensional estimates were carried out using the dimensional analysis method as proposed by P. Dirac.

4.6 Confinement and Asymptotic Freedom

The phenomenon of confinement and asymptotic freedom in a composite hadronic system has already been interpreted for hadronic metrics with considerations about the proper (hadronic) time and about the observer's coordinated (electromagnetic) time, for a detailed discussion see [4].

In summary we have that as energy varies the time elements for hadronic and electromagnetic metrics are different. Thus as energy increases the reaction time interval of the hadronic system is smaller than that of the electromagnetic action by which the hadronic system is energized and observed and therefore appears to be bound. Conversely as the energy decreases the hadronic and electromagnetic time intervals tend to equalize and the hadronic system appears pseudo free.

We now wish to make some interpretive remarks here regarding the fifth element of hadronic and electromagnetic metrics with reference to the same phenomenon.

Again we have that as the energy changes, the elements of the energy coordinate for the hadronic and electromagnetic metrics are different. The hadronic metric calibrates the energy differently from the electromagnetic metric and also the scales of the energies are different.

In summary, if the energy varies by one-tenth on the hadronic scale it is calibrated as one-millionth relative to the electromagnetic scale, that is, it is depotentiated and the system appears to the observer as weakly bound. Conversely, if the energy varies ten times on the hadronic scale it is calibrated as one million relative to the electromagnetic scale, that is, it is amplified and the hadronic system appears to the observer as strongly bound.

We do not want to push further the level of interpretation, we only observe that the two elements of hadronic metrics referring to time and energy as coordinates give consistent and coincident conclusions regarding the phenomenon of confinement and asymptotic freedom in a hadronic systems.

4.7 Interpretation of the Fifth Element of Gravitational and Electromagnetic Metrics

We now wish to make some interpretative remarks about the fifth element of the gravitational and electromagnetic metrics. Regarding the existence of both luminal and superluminal velocities in gravitation we refer the reader to Chapter 2, Section 2.2, where the problem is quoted. For the first attempt to face the problem of the 'speed of gravity' we suggest to see the fifth volume of Laplace's *Celestial Mechanics* translated into English and annotated by N. Bowditch (Boston, 1829), where the evaluation of the speed of gravitational action in the Sun-Earth-Moon system is inferred from the study of lunar libration motions.

We note that as the energy varies, the metric elements of the energy coordinate for the gravitational and electromagnetic metrics are different. Let us also remember and stress that all measurements, whether astronomical observational with various and different telescopes or experimental with various devices either in the laboratory or in orbit, still occur by electromagnetic interaction. Thus one observes gravitational phenomena with the fatally distorted view of 'electromagnetic glasses'. In fact, gravitational phenomena occur with a calibration of energy that exponentially expands energy to the sixth power. No surprise if to an 'electromagnetic observer' the gravitational systems appear to behave 'as if' there is an 'additional energy' that accelerates them in a paradoxical way while they remain visible and thus measurable electromagnetically at any frequency of electromagnetic energy itself.

Here again, we do not want to push any further the level of interpretation, except to mention that even the balance for weighing objects and even the Cavendish balance are instruments that use electromagnetic interaction to have the measurement of gravitational phenomena. The balance uses Coulombic electric repulsion between the atoms of matter making its plate and those of the weight. The Cavendish balance likewise uses the Coulombic electric repulsion between the atoms of the cable holding the dumbbell when that cable twists according to its torsion constant while the dumbbell twisting it undergoes a gravitational action.

References

[1] F. Cardone and M. Mignani. *Int. J. Modern Phys. E*, 15, 4, 911, 2008.
[2] F. Ridolfi *et al. Appl. Phys. A*, 114, 1233, 2014.
[3] R. Mignani *et al. Electron. J. Theor. Phys.*, 7, 23, 281, 2010. See also F. Cardone and R. Mignani, *Deformed Space Time*, Springer, Dordrecht, 2007. Chapter 24, Section 24.4, p. 377, Eq. 24.91.
[4] F. Cardone and R. Mignani, *Energy and Geometry*, World Scientific, Singapore, 2004, Section 10.4–10.5, p. 89–92.

Chapter 5

Five Dimensional Energy Metrics

In this chapter in order to stress the new fifth coordinate, the energy, we use for the five physical coordinates (t, x, y, z, E) the following notation $(x^0, x^1, x^2, x^3, x^5)$, already used in Section 3.2, and as in Chapter 3 we set $x^5 = l_0 E$, where the Pessa constant l_0 has the dimensions of the inverse of a linear energy density.

5.1 Lorentz Invariance and Energy as Extra Dimension

The analysis of the physical processes considered in deriving the phenomenological energy metrics for the four fundamental interactions, together experimental evidences we will discuss in the next, provides evidence for a breakdown Local Lorentz invariance (LLI), or better a violation of LLI symmetry.

But it is well known that, in general, the violation of a symmetry is the signature of the need for a wider, exact symmetry. In the case of the violation of a space time symmetry, as the Lorentz one, this is often related to the possible occurrence of higher dimensional schemes. This is indeed the case, and that energy does in fact represent an extra dimension, as already introduced in the previous chapters.

In the description of interactions by energy metrics, we saw that energy plays in fact a dual role. On one side, as more and more stressed, it constitutes a dynamical variable. On the other hand, it represents a parameter characteristic of the phenomenon considered and therefore, for a given process, it cannot be changed at will, moreover it has also the mathematical role of evolution parameter in the use of the Ricci flow method. In other words, when describing a given process, the four dimensional deformed geometry of space time, in the interaction region where the process is occurring, is 'fixed' at the situation described by those values of the metric coefficients $\{b_\mu^2(E)\}_{\mu=0,1,2,3}$ corresponding to the energy value of the process considered. Namely, a fixed value of E determines the space time structure of the interaction region at that given energy. In this respect, therefore, the energy of the process is to be considered as a geometrical quantity intimately related to the very geometrical structure of the physical world. In other words, from a geometrical point of view, all goes on as if we were actually working on 'slices' or better sections of a five dimensional space, in which the extra dimension is just represented by the energy.

The simplest way to take account of and to make explicit the double role of energy in deformed space time is assuming that E represents a fifth metric dimension, on the same footing of space and time, and therefore to embed the four dimensional deformed Minkowski space $\widetilde{M}_4(E)$ in a five dimensional Riemann space R_5, as already done in the previous chapters, but with motivations other than the present. By the way all this seems to lead to the building up of a sort of pseudo 'Kaluza Klein like' scheme, with energy as fifth dimension. But this similitude is only apparent and misleading, since the Deformed Space Time theory is not compactified and the cylindricity condition is not implemented, and even reversed. Actually, as we shall see the metric tensor $\overset{(5)}{g}_{\mathrm{AB}}(E)$ depends only on the fifth coordinate E.

5.2 Deformed Space Time and Riemannian Space

We assume then a five dimensional space-time R_5 endowed with the energy metric $\overset{(5)}{g}_{\mathrm{AB}}(E)$, in the following of this chapter capital Latin indices take values in the range (0, 1, 2, 3, 5), with index 5 labelling the fifth dimension. For the sake of mathematical completeness we put the metric also in the following form by using the

Kronecker delta:

$$\begin{aligned} g^{(5)}_{\mathrm{AB}}(E) &= \mathrm{diag}(b_0^2(E), -b_1^2(E), -b_2^2(E), -b_3^2(E), b_5^2(E)) \\ &= \delta_{\mathrm{AB}}(b_0^2(E)\delta_{A0} - b_1^2(E)\delta_{A1} - b_2^2(E)\delta_{A2} \\ &\quad - b_3^2(E)\delta_{A3}) + b_5^2(E)\delta_{A5}. \end{aligned} \tag{5.1}$$

It follows from Eq. (5.1) that E, which is an independent non-metric variable in four dimensional deformed space time, becomes a *metric* coordinate in R_5. Then, whereas $g_{\mu\nu}\ (E)$ is a deformed, Minkowskian metric tensor, $\boldsymbol{g^{(5)}_{AB}(E)}$ is a genuine Riemannian metric tensor.

Therefore, the infinitesimal interval $ds^2_{(5)}$ of R_5 is given by:

$$\begin{aligned} ds^2_{(5)} &\equiv g^{(5)}_{\mathrm{AB}}(E)dx^A dx^B \\ &= b_0^2(E)(dx^0)^2 - b_1^2(E)(dx^1)^2 - b_2^2(E)(dx^2)^2 \\ &\quad - b_3^2(E)(dx^3)^2 + b_5^2(E)(dx^5)^2 \\ &= b_0^2(E)c^2(dt)^2 - b_1^2(E)(dx^1)^2 - b_2^2(E)(dx^2)^2 \\ &\quad - b_3^2(E)(dx^3)^2 + b_5^2(E)l_0^2(dE)^2, \end{aligned} \tag{5.2}$$

where we have put

$$x^5 = l_0 E, \quad l_0 > 0. \tag{5.3}$$

The Pessa constant l_0 provides, as usual, the dimensional conversion energy $\to$ length and, as reminded, it has therefore the dimensions of the inverse of a linear energy density.

The coefficients $b_\mu^2(E)$ of the metric of $\tilde{M}_4(E)$ can be therefore expressed as

$$\left\{b_\mu\left(\frac{E}{E_0}\right)\right\} \equiv \left\{b_\mu\left(\frac{x^5}{x_0^5}\right)\right\} = \left\{b_\mu(x^5)\right\}, \quad \forall\mu = 0, 1, 2, 3, \tag{5.4}$$

where we put

$$x_0^5 = l_0 E_0. \tag{5.5}$$

As to the fifth metric coefficient, one assumes that it too is a function of the energy only: $b_5 = b_5(E) = b_5(x^5)$. Although, in principle, nothing prevents from assuming that, in general, b_5 may depend also on space time coordinates x^μ: $b_5 = b_5(x^\mu, x^5)$. Unlike the other metric elements, it may be $b_5(E) \gtrless 0$. Therefore, a priori, the energy dimension may have either a timelike or a spacelike signature in R_5, depending on $\mathrm{sgn}(b_5(E)) = \pm 1$.

There is therefore an uncertainty in the sign of the energy metric coefficient $b_5^2(E)$. In particular, it cannot be excluded a priori that the signature of E can change. This occurs whenever the function $b_5(E)$ does vanish for some energy values. As a consequence, in correspondence to the energy values which are zeros of $b_5(E)$, the metric $g_{\mathrm{AB}}^{(5)}(E)$ could be degenerate. But the above considerations are only academic just for the sake of completeness since it is now known that $b_5(E)$ is ever different from zero as we saw from Ricci flow equations in Chapter 4. Anyway the problem remains of $\mathrm{sgn}(b_5(E))$ if timelike or spacelike, almost at present knowledge.

5.3 Five Dimensional Energy Metric Overview

Turning to the representation, energy takes on the role of coordinate. Energy is an additional measurable and extended real physical dimension, thus endowed with measurable physical dimensions. The four pentadimensional metrics associated with the four fundamental interactions are defined in a space time energy manifold endowed with global length-dimensional coordinates so that the elements of the metric tensor turn out to be dimensionless. The first coordinate represents time through the product, that is, the product of time by the velocity u which is the maximum relativistically invariant causal velocity corresponding to each interaction. The fifth coordinate represents the energy through the now well known product

$$x_5 = l_0 E.$$

The intermediate spatial coordinates have the dimension of a length.

As seen from their pentadimensional expressions we can now fix the following two points about the behaviour of the fifth metric element according to each interaction.

1. For hadronic and gravitational interactions it is a power of energy measured in units of its threshold energy and does not respect any type of threshold, so this element of the metric acts even when the

metric is flat, even if for energy values well below the threshold it approaches to 1.

2. For leptonic and electromagnetic interactions it is a constant and therefore indifferent to whether the metric is flat or not flat, or rather deformed.

Here we repeat the tentative hypothesis 'ansatz' (in German after Einstein) that the fifth parameter of the metric for each interaction is the calibration of the energy for that interaction. In fact it describes for the energy coordinate how it is modified in the metric of the interaction itself, i.e. in simple words, how '*in each interaction energy measures energy*'. Again we remind that the reference energy for each phenomenon is that measured with instruments that use electromagnetic interaction in conditions of flat space time and the validity of Hamilton's theorem for the conservation of total energy.

Therefore:

'The electromagnetic metric and the electromagnetic interaction constitute the fixed point of reference in all measurements of phenomena also when governed by other interactions with their relative metrics'.

The result that the fifth element of the metric, corresponding to the coordinate energy, has a functional dependence on a power of the energy, had already been hypothesized previously. Each pentadimensional energy metric has it's own energy threshold and we already divided them in: over-Minkowskian if the deformed metric becomes flat while energy E decreases from values above the threshold, sub-Minkowskian if the deformed metric becomes flat while energy E increases from values below the threshold. We resume these behaviours in a qualitative way in Figs. 5.1 and 5.2.

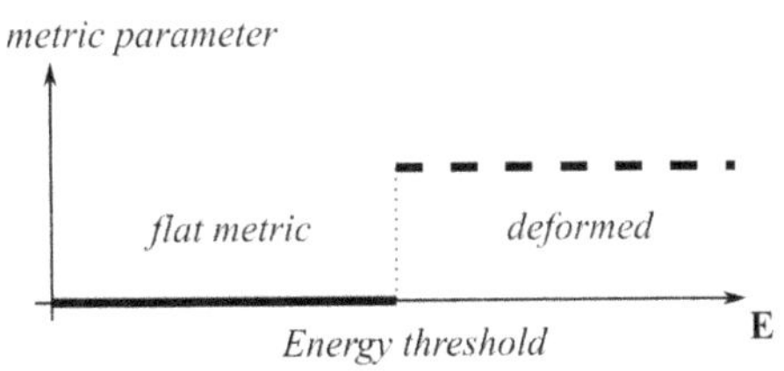

Fig. 5.1. Over-Minkowskian metric, gravitational and hadronic.

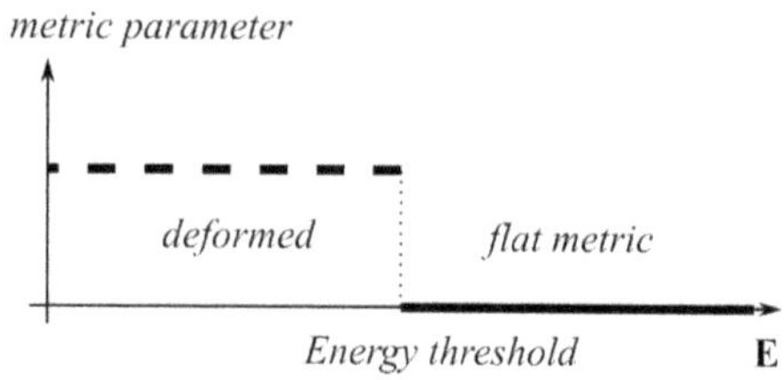

Fig. 5.2. Sub-Minkowskian metric, e.m. and leptonic.

So, we said that over-Minkowskian metric interactions, such as hadronic and gravitational, have a variable energy calibration with the energy itself, regardless of whether the metric is flat or deformed, and on the other hand, sub-Minkowskian metric interactions, such as electromagnetic and leptonic, always have the same energy calibration regardless of the value of the energy itself. We now definitively set this interaction calibration parameter $k_{em} = 1$ for the electromagnetic interaction as a convention and for convenience.

The results achieved by the Ricci flow have allowed to design, build and test devices capable of exploiting the behaviour of the fifth element of the metrics, in particular hadronic and leptonic, to obtain the production of electric charges directly from the nuclear metamorphosis of the matter as it will be discussed in next chapters. At last, when applied to nuclear metamorphosis, the expression of the fifth element of the hadronic metric obtained by Ricci flow works better then that one obtained by Killing equations, perhaps due to the heavy hypotheses chosen for deducing the solutions of the Killing equations, as illustrated in Section 3.2.

We underline that the results relating to the fifth element of the metric can be summarized for all interactions with the following expression of the functional dependence on energy:

$$b_5^2(E) = E^r, r \in Q + \{0\}.$$

This expression, which provides the functional form of the dependency from energy, had already been hypothesized previously in the context of the 12 classes of solutions of Einstein's field equations in vacuum for penta dimensional metrics in deformed space-time, as shown in brief in Section 3.1. It is therefore a further result to have verified by the Ricci flow method that this previous hypothesis is correct. We summarize in the following figures the results already exposed for each interaction. Below, for each interaction, we report the forms matching the experiments. These forms contain data and reporting once more the values of the thresholds $E_{0\mathrm{int}}$, in this way it became also possible to reprocess these data to display the trend of $b^2(E)$ in numerically reliable graphs.

Transcription of the electromagnetic metric (see Fig. (5.3)):

Threshold energy	$E_{0em} = 4.5\,\mu eV$
$b_0^2(E) = 1$	
$b_1^2 = b_2^2 = b_3^2(E) = \left(\frac{E}{E_{0em}}\right)^{1/3},$	$E < E_{0em}$
$b_1^2 = b_2^2 = b_3^2(E) = 1,$	$E > E_{0em}$
$b_5^2(E) = k_{em},$	$k_{em} > 0$ constant
$b_5^2(E) = b_0^2(E) = 1,$	For $k_{e.m.} = 1$ constant self-calibrated energy

For convention and convenience we have chosen $k_{e.m.}$ coinciding with 1 since, as said before, the electromagnetic interaction is the paradigm of all our phenomenological and experimental measurements as all the instruments at our disposal to date work with it. So it is with this meaning that energy is considered self-calibrated for electromagnetic interaction.

We get the values of the metrics before the threshold, on the threshold and after the threshold:

$$E < E_{0em} \begin{cases} b_0^2(E) = 1 \\ b_1^2(E) = b_2^2(E) = b_3^2(E) = \left(\frac{E}{E_{0em}}\right)^{1/3} \\ b_5^2(E) = k_{em} \end{cases}$$

$$E = E_{0em} \begin{cases} b_0^2(E_{0em}) = 1 \\ b_1^2(E_{0em}) = b_2^2(E_{0em}) = b_3^2(E_{0em}) = 1 \\ b_5^2(E) = k_{em} \end{cases}$$

$$E > E_{0em} \begin{cases} b_0^2(E) = 1 \\ b_1^2(E) = b_2^2(E) = b_3^2(E) = 1 \\ b_5^2(E) = k_{em} \end{cases}$$

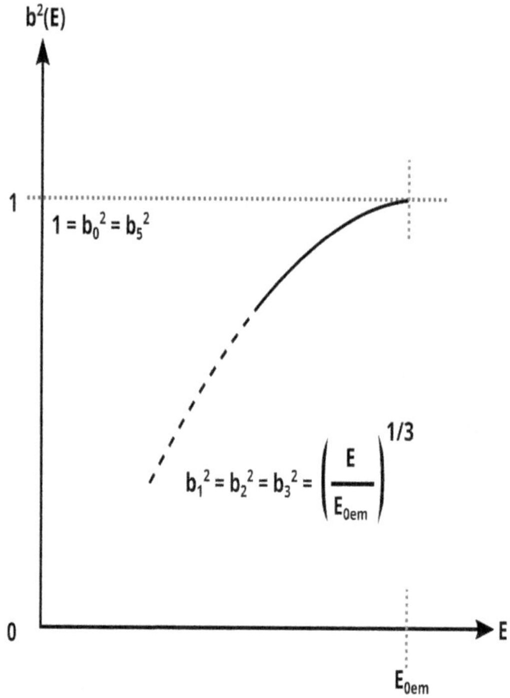

Fig. 5.3. Electromagnetic metric representation.

Transcription of the leptonic metric (see Fig. (5.4)):

Threshold energy	$E_{0\text{lep}} = 80.4\,\text{GeV}$
$b_0^2(E) = 1$	
$b_1^2(E) = b_2^2(E) = b_3^2(E) = (E/E_{0\text{lep}})^{1/3}$,	$E \leq E_{0\text{lep}}$
$b_1^2(E) = b_2^2(E) = b_3^2(E) = 1$,	$E > E_{0\text{lep}}$
$b_5^2(E) = k_{\text{lep}} > 0$	k_{lep} constant
$b_5^2(E) = b_0^2(E) = 1$	valid for constant calibrated energy as for the e.m. metric: $k_{\text{lep}} = k_{em} = 1$

We observe that, the remaining conditions in the case given, the range where this last condition is valid it is the entire energy axis.

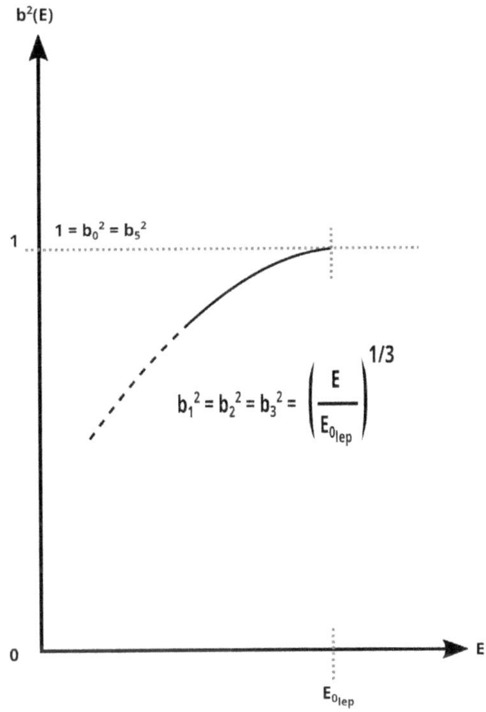

Fig. 5.4. Leptonic metric representation.

Transcription of the gravitational metric (see Fig. (5.5)):

Threshold energy	$E_{0\text{grav}} = 20.2\,\mu eV$
$b_0^2(E) = 1,$	$E \leq E_{0\text{grav}}$
$b_1^2(E) = b_2^2(E) = b_3^2(E) = 1$	$E < E_{0\text{grav}}$
$b_5^2(E) = K_{\text{grav}} > 0$ *constant*	$E < E_{0\text{grav}}$
$b_5^2(E) = b_0^2(E) = 1,$	for $K_{gra} = 1$ constant calibrated energy as for e.m. metric

$$b_3^2(E) = b_0^2(E) = \left(1 + \frac{E}{E_{0\text{grav}}}\right)^2, \quad E \geqslant E_{0\text{grav}}$$

$$b_1^2(E) = b_2^2(E) = 1$$

$$b_5^2(E) = \left(1 + \frac{E}{E_{0\text{grav}}}\right)^6, \qquad E \geq E_{0\text{grav}}$$

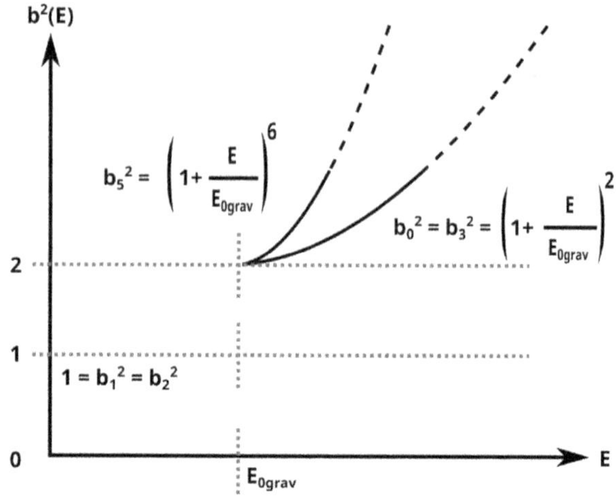

Fig. 5.5. Gravitational metric representation.

Transcription of the hadronic metric:

Threshold energy	$E_{0\text{had}} = 367.5\,\text{GeV}$
$b_0^2(E) = 1,$	$E \leq E_{0\text{had}}$
$b_1^2(E) = \left(\frac{\sqrt{2}}{5}\right)^2 = 0.08$	$b_2^2(E) = \left(\frac{2}{5}\right)^2 = 0.16$ *anisotropy*
$b_3^2(E) = 1$	$E < E_{0\text{had}}$
$b_5^2(E) = K_{\text{had}} > 0$	K_{had} constant
$b_5^2(E) = b_0^2(E) = 1$	for $K_{\text{had}} = 1$ constant calibrated energy as for e.m. metric
$b_0^2(E) = \left(\frac{E}{E_{0\text{had}}}\right)^2,$	$E \geqslant E_{0\text{had}}$
$b_1^2(E) = \left(\frac{\sqrt{2}}{5}\right)^2$	$b_2^2(E) = \left(\frac{2}{5}\right)^2$ anisotropy
$b_3^2(E) = b_0^2(E) = \left(\frac{E}{E_{0\text{had}}}\right)^2,$	$E \geqslant E_{0\text{had}}$
$b_5^2(E) = \left(\frac{E}{E_{0\text{had}}}\right)^6,$	$E \geqslant E_{0\text{had}}$

$b_0^2(E)$ is equal to 1 up to the threshold where it continues with $(E/E_{0\text{had}})^2$;

$b_1^2(E), b_2^2(E)$ their constant values are given regardless of the threshold and therefore for every value of the energy E;

$b_3^2(E)$ is equal to 1 before the threshold, is equal to 1 at the threshold and $(E/E_{0\text{had}})^2$ after the threshold. Ultimately it is $b_0^2(E) = b_3^2(E)$;

$b_5^2(E)$ is equal to $(E/E_{0\text{had}})^6$ on the threshold and for energy values greater than the threshold. For energy values lower than the threshold we assumed conventionally a value equal to 1, in accordance with $K_{0\text{had}}$. In this sense we put the convention that the energy gauge does not act below the hadronic threshold.

Part III

Energy Geometry

Chapter 6

Neutron Emissions

In this chapter we examine the experiments that produced neutron emissions from reactions predicted according to the Deformed Space Time (DST) theory.

6.1 Neutron Impulses in the Time

The first experiments that produced neutron emissions from DST reactions were conducted from 2004, and published from 2007, on ultrapure deionized water with a resistivity of $18.2\,\mathrm{M\Omega \cdot cm}$, on aqueous solutions of Ferric Chloride and Ferric Nitrate, Aluminium Chloride, and Lithium Chloride.

Equipment and materials used were the following: a power supply connected to an ultrasonic generator consisting of a sealed column of piezoelectrics; a conical trunk transmitter made of AISI 304, called the 'Sonotrode' and borosilicate glass beakers. All solutions were in ultrapure deionized water with a resistivity of $18.2\,\mathrm{M\Omega \cdot cm}$. Ultrasound, with a frequency of 20 kHz and a power of 130 W was transmitted to the Sonotrode and applied for 90 min to 250 ml solutions of $FeCl_3$, $AlCl_3$, and LiCl, and for 120 min to a 500 ml solution of $Fe(NO_3)_3$.

The ultrasonic transmitting sonotrode was immersed into the solutions, perpendicularly to the free surface, positioned exactly at the center of their container. The distance from the tip of the sonotrode to the bottom of the container was 4 cm for the $FeCl_3$,

$AlCl_3$, and LiCl solutions, while it was 5 cm for the $Fe(NO_3)_3$ solution. These distances were chosen to achieve the maximum possible 'energy density' in the solution. During the experiment, the ultrasonic transmitter was cooled using a compressed air system. During the experimental trials neutron passive detectors were installed, called Defender and CR-39 PADC, respectively. In a subsequent phase of the experiment, BF3 (Boron Trifluoride) detectors were employed. Measurement of alpha, beta and gamma radiation were also performed using Geiger Gamma Scout detector and NaI(Tl) scintillation detectors. Three Defender detectors were positioned vertically, parallel to the axis of the sonotrode and the cavitation container. Here, the term 'cavitation' refers to the application of ultrasound to solutions, i.e. substances in the liquid phase.

A schematic representation of the experimental setup is shown in Fig. 6.1.

While no neutron emission was detected or measured in the cavitation trials conducted on ultrapure deionized H_2O and solutions of $AlCl_3$ and LiCl, neutron emission was measured in the experiments performed on $FeCl_3$ and $Fe(NO_3)_3$ solutions. For ultrapure deionized H_2O and for the solutions of $AlCl_3$ and LiCl, the neutron dose measurement was compatible with the environmental background: 20 nSv. For the $FeCl_3$ and $Fe(NO_3)_3$ solutions, on the contrary,

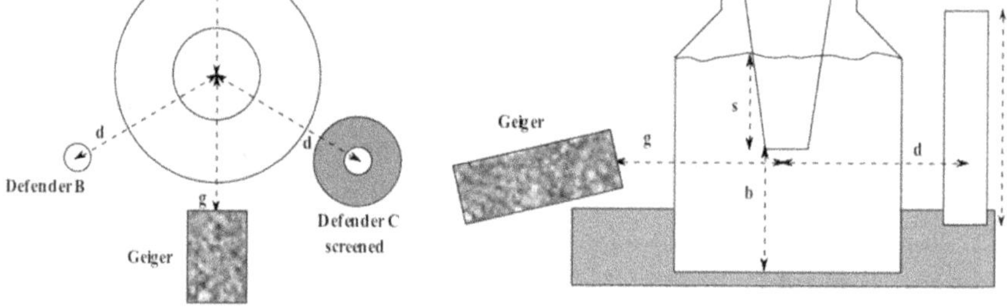

Fig. 6.1. Top view and side view of the experimental setup. d = 7 cm, g = 10 cm, s = 4 cm, b = 5 cm. The figure shows that between the cavitation zone, the neutron detectors and the Geiger counter there were 3.5 cm of solution, 2 mm thick borosilicate glass and a few cm of air.

the neutron dose was not compatible with the environmental background [1].

The subsequent Figs. 6.2 and 6.3 show the quantitative trend of neutron emission as a function of cavitation time for the $FeCl_3$ and $Fe(NO_3)_3$ solutions.

The graphs demonstrate that the neutron emission induced by cavitation reveals the existence of a clear threshold level as a function of time. Notably, the measured levels of alpha, beta, and gamma radiation were compatible with the environmental background in all experiments and for the first 90 minutes or so. Therefore, the increase in neutron emission as a function of cavitation time, which becomes evident in the last 30 minutes of the experiment on iron salt solutions, can be seen as an initial experimental confirmation of the phenomenological model proposed by the Deformed Space-Time theory. A subsequent step involved replicating the experiment on 250 ml of a $FeCl_3$ solution with a concentration of 1000 ppm (1 g/L) under the same experimental conditions described above, but employing a BF_3

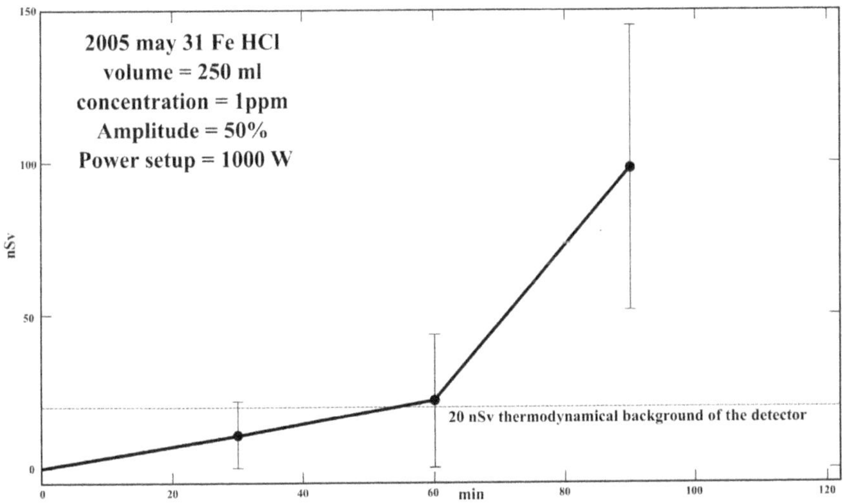

Fig. 6.2. Neutron dose (nSv) as a function of cavitation time for the $FeCl_3$ solution. The horizontal line represents the background level, given by the sum of the thermodynamic instability of the detectors and the environmental background, and is equal to 20 nSv. It is evident that in the last 30 min of cavitation, the neutron dose is significantly higher than the background, approximately $\approx$ 100 nSv, about 5 times the background, more precisely 98.5 ± 45 nSv.

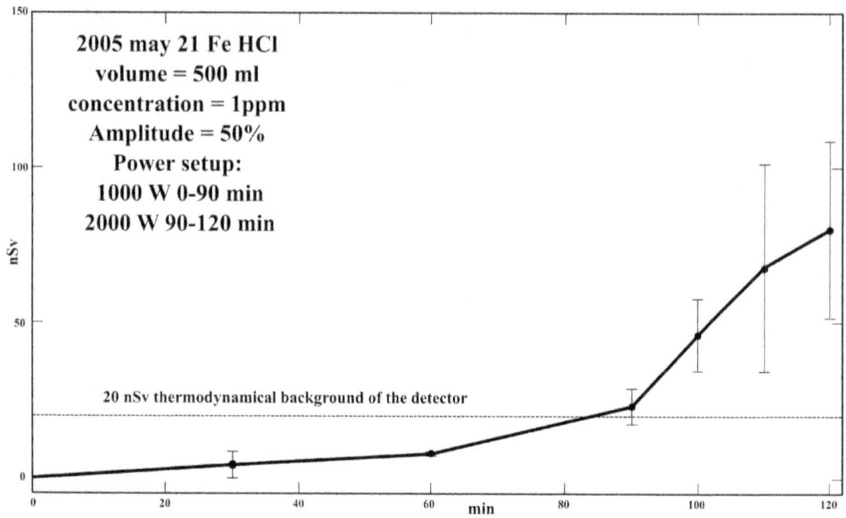

Fig. 6.3. Neutron dose (nSv) as a function of cavitation time for the $Fe(NO_3)_3$ solution. The horizontal line represents the background level, given by the sum of the thermodynamic instability of the detectors and the environmental background, and is equal to 20 nSv. It is evident that in the last 30 minutes of cavitation, the neutron dose is significantly higher than the background, approximately ≈ 100 nSv, about 5 times the background, more precisely 76.0 ± 45 nSv.

neutron detector (Weldom Medical 2222A) [2]. Ionizing radiation emissions, particularly gamma rays, were also re-investigated using a UMo LB 1236 detector. The neutron detector was calibrated using a ^{241}Am-Be source to provide a contact dose rate of $1.5 \pm 0.2\,\mu$Sv/h.

Figure 6.4 shows the setup of the experimental apparatus.

Neutron emission was observed, as shown in Fig. 6.5.

The graph is divided into two parts by a vertical line placed after approximately 90 min.

Before the line represents the time period during which cavitation was ongoing; after the line, between 90 and 180 min, represents the time period following the end of cavitation. In this second time period, neutron emission persists. Throughout the entire time, neutron emission in pulses is evident, not constant over time. The fact that the distribution of the pulses, upon careful analysis, does not exhibit a Gaussian shape indicates that they are not identical pulses emitted at different times in various directions, but rather entirely distinct pulses. Several cavitation tests were performed; in all tests,

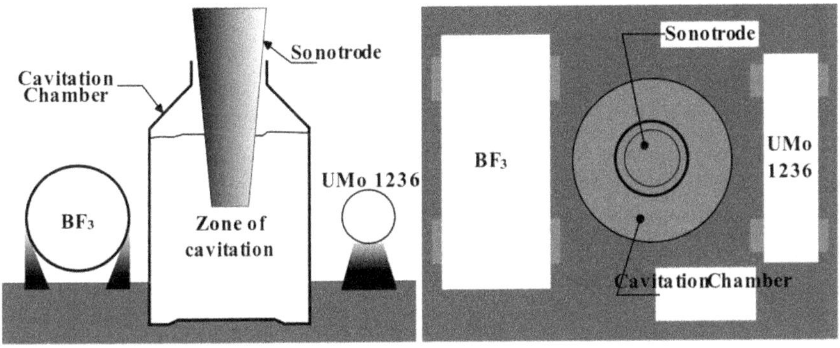

Fig. 6.4. Vertical and horizontal representations of the apparatus used for the cavitation of $FeCl_3$ in solution.

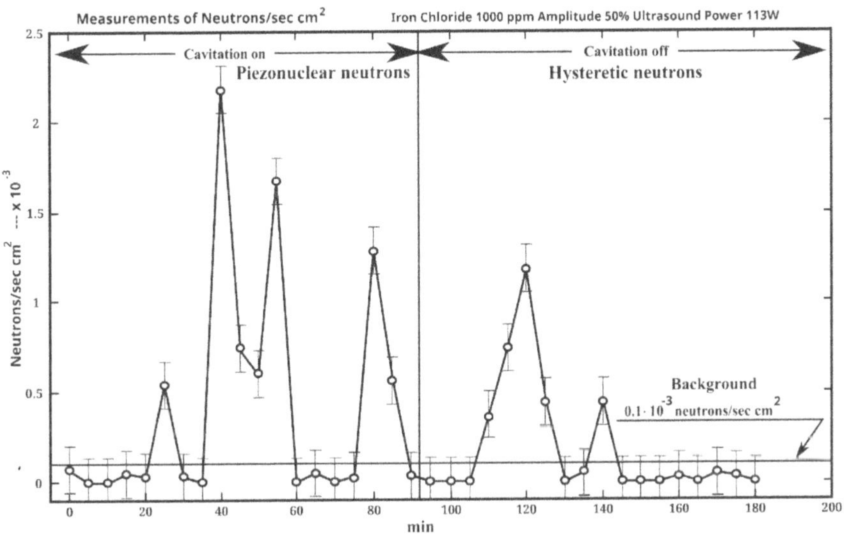

Fig. 6.5. Graph of neutron pulses obtained during the cavitation of 250 ml of a 1000 ppm $FeCl_3$ solution. The x-axis represents time in minutes, and the y-axis represents the neutron flux in $(n\cdot s^{-1}\cdot cm^2)\cdot 10^{-3}$.

the first appearance of the neutron burst occurred 40–50 min after the start of cavitation. This further confirms the existence of a threshold level in time, and therefore an energy threshold level, in DST neutron emission.

The neutron bursts emitted after the end of cavitation, approximately 20 min later, represent 'hysteretic' pulses emitted after a latency period, delayed pulses, similar to what occurs in nuclear reactors due to graphite. In this case as well, no gamma radioactivity emission was detected above the background. Furthermore, it is also noteworthy that in the solutions, after the experimental trials, the aluminium level was found to have increased by about a factor of seven compared to the initial level: from 3.99 to 27.70 ppb (nanograms/gram). Excluding the possibility of transmutations, due to the absence of radioactivity, a possible explanation could lie in the following DST fission reaction:

$$Fe_{26}^{30} \rightarrow 2Al_{13}^{14} + 2n. \tag{6.1}$$

Finally, it is significant that the emissive phenomenon occurred in the case of iron salts, but not in the case of aluminium and lithium salts, that is, in the case of one of the elements that show the highest value of binding energy per nucleon, as evident from the Aston, Bohr, and Wheeler curve shown in Fig. 6.6.

The next step in the experimental trials was the transition to cavitation of solid materials with a high iron content, that is, iron alloys and steel.

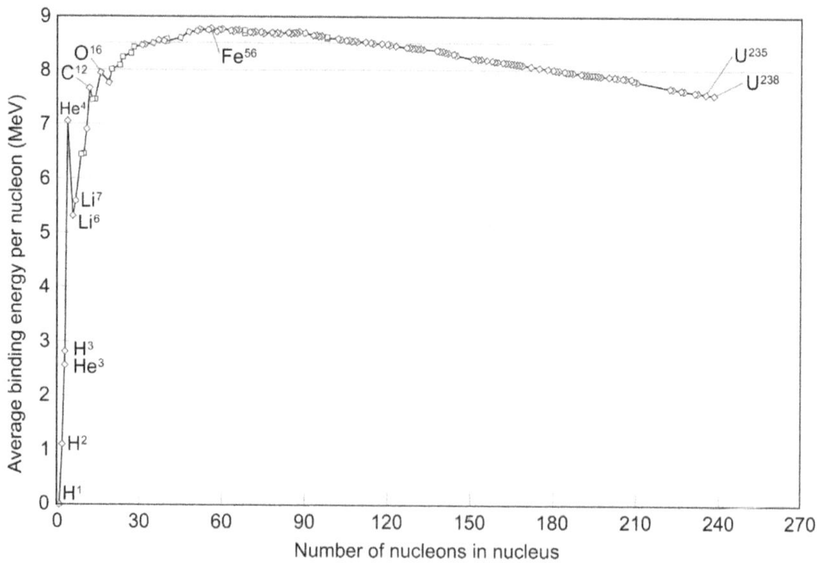

Fig. 6.6. Average binding energy per nucleon for a selection of nuclides.

6.2 Neutron Impulses in the Energy and Space

In the transition to sonication of solid materials, specifically cylindrical bars of Ferrite and AISI 304, for the stimulation of DST reactions and the associated production of neutrons, there are two main advantages in terms of power: 1) the predominant presence of iron, a key component of the alloys; 2) the greater density of the material, about 7.8 for steels compared to about 1 for aqueous solutions.

The first advantage allows for the exploitation of the DST reactions of iron, enhanced by the high binding energy per nucleon of its isotopes, see Fig. 6.6; the second advantage makes it possible to significantly increase the energy density obtainable during cavitation, which benefits from the higher density of the material. Both phenomena can lead to a higher neutron emission and an earlier onset of the neutron manifestation. The first experiment was conducted in 2009 at the Startec laboratories in Milan. Cylindrical bars of Ferrite and steel bars, approximately 20 cm in length and 2 cm in diameter, were sonicated. An appropriately designed metal support apparatus allowed for perfect contact between the sonotrode and the bars, while a hollow dielectric material cylinder carefully maintained the bars in a fully vertical alignment. At the bottom, opposite the sonotrode, a reflector made of the same material as the sonotrode was placed.

The experimental conditions were similar to those adopted in all the previous tests: a power supply connected to an ultrasonic generator consisting of a piezoelectric box; the ultrasound, with a frequency of 20 kHz and a transferred power of 19 W, was directed to a truncated-cone transmitter made of AISI 304, the 'Sonotrode' and the vibration amplitude of the sonotrode tip was 15 micrometers. Ultrasound was applied for 90 minutes. During the experiment, the ultrasound transmitter was cooled using a compressed air system.

A diagram of the apparatus is shown in the following Fig. 6.7.

Neutron measurements were performed using two types of detectors. The first was an HDS-100GN Neutron Counter by Mirion, equipped with a LiI(Eu) scintillation neutron detector and a CsI(Tl) crystal for gamma rays. The second was a MICROSPEC2 BTI neutron probe, consisting of a multichannel analyzer connected to a ^{3}He neutron counter for thermal neutrons up to 0.8 MeV and a liquid scintillator for neutrons with energies range mainly from 0.5 to 5 MeV, but possible extension up to 18 MeV for fluency measure.

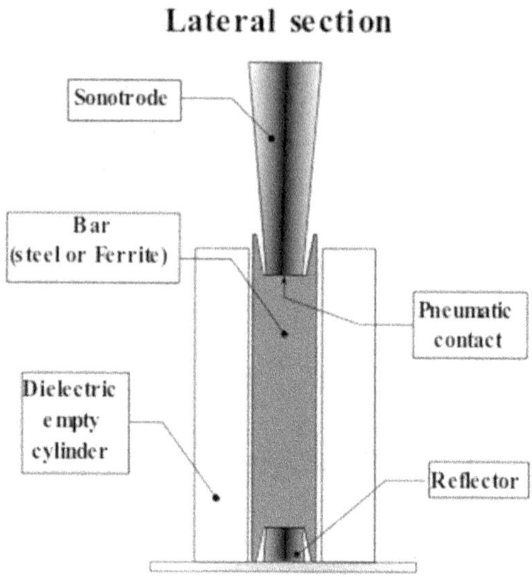

Fig. 6.7. Apparatus schema.

The counters were placed 20 cm away from the outer surface of the hollow dielectric cylinder that surrounded and secured the sonicated bar. During the sonications performed on both Ferrite bars and Steel bars, two tests per bar, neutron emissions in sudden bursts were detected, with the intensity of these bursts exceeding the maximum ambient background level by 25%.

Another significant experimental observation regarding neutron emission was the time of appearance of the first neutron pulse: in all tests, it occurred after only about five minutes from the start of sonication. This represented a marked difference compared to the time of neutron appearance during previous cavitations of ferrous salt solutions, where the first burst was detected after 40–50 min from the start of cavitation [3].

Figure 6.8 shows the graphs of the neutron spectra. The neutron energy ranged from 0 to 3 MeV.

No gamma radiation above the background level was detected by the HDS-100GN Mirion detector.

Further significant evidence was the appearance, after sonication, of small circular cavities on the external lateral surface of the bars, with diameters ranging between 2 and 3 mm, as shown in Fig. 6.9.

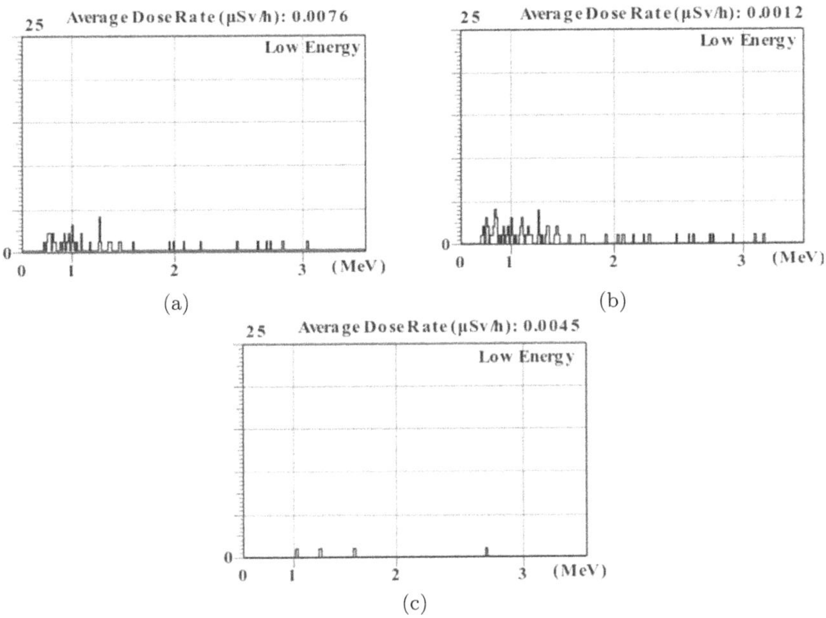

Fig. 6.8. (a) Neutron spectrum from the Ferrite bar. (b) Neutron spectrum from the Steel bar. (c) Neutron spectrum of the background. On the y-axis, the number of recorded pulses; on the x-axis, the neutron energy in MeV.

These circular cavities might have been generated by the escape of neutrons during sonication. They are indeed entirely similar to the cavities produced by neutron escape in the cladding of nuclear fuel elements used in nuclear reactors.

Regarding the measured neutrons, see Fig. 6.8, one might remain unsatisfied with the low count levels, despite the clear difference between the emission measurements from the sonicated bars and the background measurement.

The explanation for this could lie in two main reasons: firstly, as already mentioned, DST neutron emission is not constant but occurs in bursts; secondly, the efficiency of active detectors, such as those used, is very low. This second limitation, caused by the measurement dead time, means that rapid and sudden bursts of neutrons might be missed during counting. Moreover, neutron bursts, especially if caused by DST reactions, may not be emitted isotropically, following a 4π steradian geometry, but rather concentrated in a preferential direction. This is suggested by Fig. 6.9, which shows that the traces

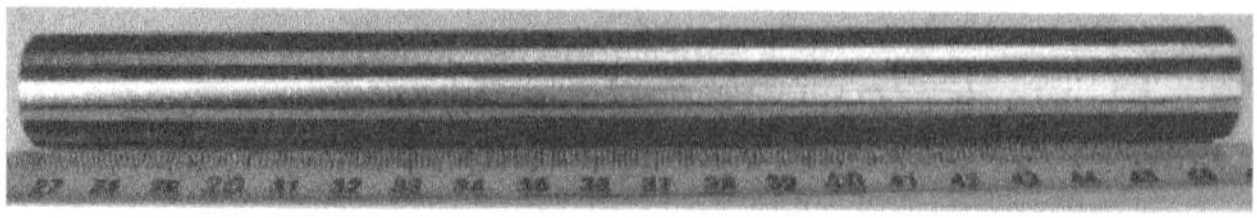

(a)

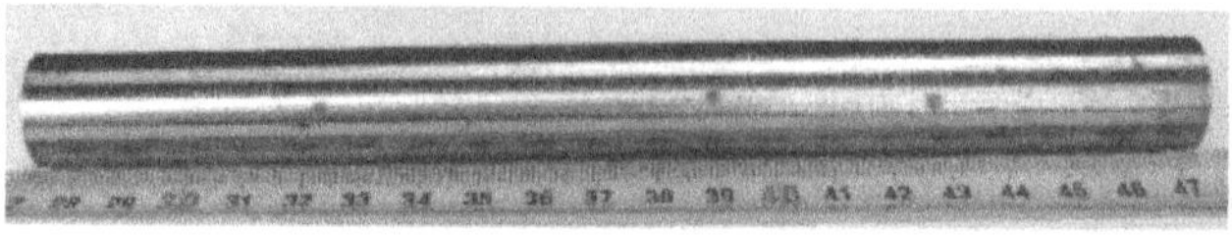

(b)

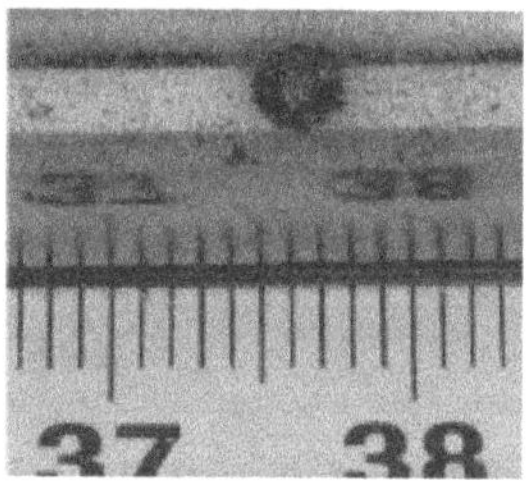

(c)

Fig. 6.9. (a) Not sonicated bar. (b) Sonicated bar. (c) Detail of one of the damages on the sonicated bar.

left by escaping neutrons are unevenly distributed across the surface of the bars [4].

Another important observation comes from analyses conducted using SEM-EDX techniques (Scanning Electron Microscopy — X-ray Emission); these revealed significant negative variations in the concentration of Iron in areas damaged by neutron escape compared to undamaged areas, as well as substantial increases in light elements Oxygen, Carbon and the presence of new, unexpected elements. Specifically, the decrease in Iron concentration was approximately −47.5%, while the combined increases in Oxygen and Carbon were around +46.7%. It can therefore be hypothesized that Iron transforms into other lighter elements through mechanisms which, given the consistent absence of radioactivity, have nothing to do with classical nuclear processes according to DST theory.

The following DST reactions could potentially be proposed, which also explain the detected neutron emission.

$$^{56}Fe^{30}_{26} \rightarrow 3^{12}C^{6}_{6} + ^{16}O^{8}_{8} + 4n \tag{6.2}$$

$$^{56}Fe^{30}_{26} \rightarrow 2^{12}C^{6}_{6} + ^{16}O^{8}_{8} + 3^{4}\alpha^{2}_{2} + 4n. \tag{6.3}$$

A subsequent experiment was conducted between 2012 and 2013, sonicating a steel bar made of AISI 304, with a mass of 180 g, 20 cm long, and a diameter of 2 cm, using ultrasound at a frequency of 20 kHz and a power of 330 W, for 180 seconds. The sonication was performed using the same machine employed in the previous sonication experiments on Ferrite and Steel bars.

Neutrons detectors used included the active MICROSPEC2 system and the HDS Mirion system to monitor the possible emission of ionizing radiation in the form of gamma rays.

Additionally, 16 passive detectors called CR-39 AB were uniformly placed around the sonicated steel bar. The CR-39 AB detectors were immersed in finely divided boric acid (H_3BO_3, or AB) powder and had been previously tested and calibrated at the ENEA Research Center in Rome-Cesano (Casaccia) using three neutron sources: an ^{241}Am-Be source, the TRIGA nuclear reactor (radial channel), and the TAPIRO nuclear reactor [5]. The configuration of the entire sonotrode-steel bar (AISI 304) system was essentially identical to that used in the previous experiments, as described in Fig. 6.7, with the dielectric material consisting of PTFE (Teflon). During the experiment, as in previous cases, the ultrasonic generator system was cooled using a compressed air system. In this experimental trial as well, neutron emission was detected. Figure 6.10 presents the neutron emission spectra and the measured background, obtained using the MICROSPEC2 system [6].

Figure 6.11 Neutron emission spectra detected during sonication, shown in greater detail for the energy ranges 0 to 0.5 MeV and 0 to 4 MeV, obtained using the MICROSPEC2 system.

At a first naive glance Fig. 6.10 could seem containing a contradiction. But it is only apparent since it is due to the scale factor of the vertical axes, having the same number of divisions, once considered that the typical value of the macroscopical cross section for detection is about in the range $0.2 - 0.3\,\text{cm}^2$.

The sixteen passive PADC CR-39 AB detectors were then processed. At the end of the sonication, each individual detector was retrieved, specifically treated to highlight neutron tracks, observed, and photographed using an optical microscope (100x). For each photo of all the sixteen CR-39 AB detectors, measuring 10×10 cm, neutron track counting was performed by dividing each photo into 400 squares using a grid with 0.5 cm sides. The same procedure was applied to

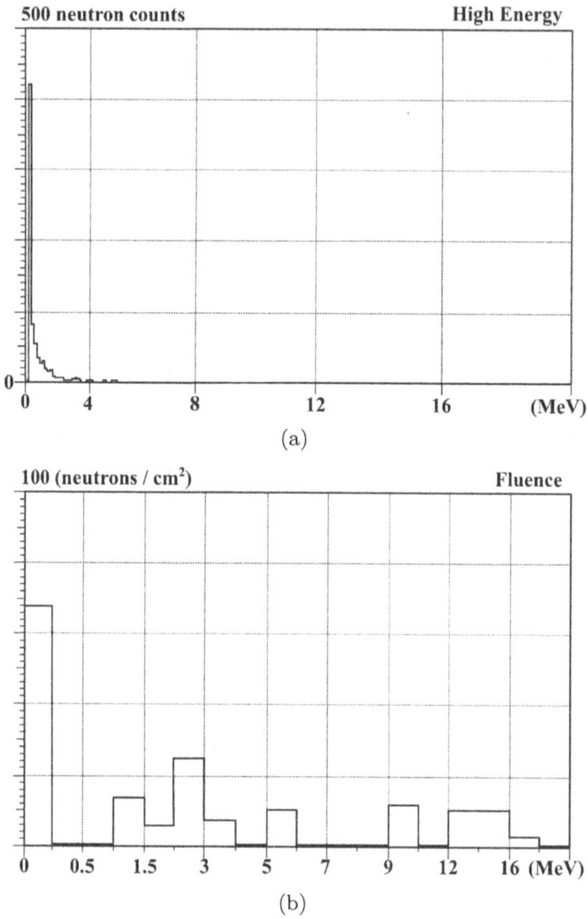

Fig. 6.10. MICROSPEC2 Detector. Energy spectrum of neutrons detected during the sonication of AISI 304. (a) The neutron counts are shown on the y-axis, and energy (MeV) on the x-axis. (b) The neutron fluence (n/cm^2) is shown on the y-axis, and energy (MeV) on the x-axis. (c) Background: Neutron counts are shown on the y-axis, and energy (MeV) on the x-axis, no counts above 3 MeV.

the blanks, four CR-39 AB detectors. This identical method was also used for the images obtained from the previously mentioned reference neutron sources [8, 9].

For neutron tracks, a binary yes/no criterion was employed for each square of the grid on every detector: presence of tracks (yes) or absence of tracks (no). The results are shown in Fig. 6.12.

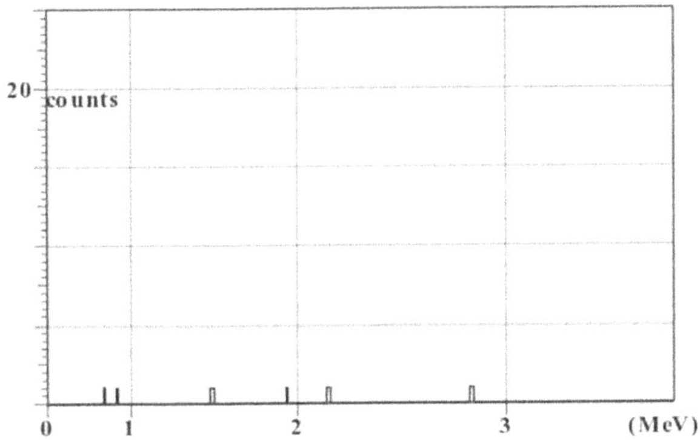

(c)

Fig. 6.10. (*Continued*).

In the horizontal view, at the center, the AISI 304 steel bar is depicted, surrounded by the hollow PTFE cylinder. Surrounding this, arranged in a circular pattern, are the photographs of the sixteen CR-39 AB detectors. In the bottom-right panel, the photographs of the four CR-39 AB detectors with the background measurements are displayed. In the two panels further below: on the left, photographs of the neutron spectra obtained during sonication and background measurements using the MICROSPEC2 detector; on the right, the photographs of the CR-39 AB detectors showing the images produced by the reference neutron sources.

In panels I, II, III, and IV, the neutron measurement results from the CR-39 AB detectors are presented for each individual $\pi/4$ quadrant (90°).

Firstly, it can be stated that, as in previous cases, the HDS Mirion detector did not detect gamma radioactivity emissions exceeding the background level. Additionally, the HDS Mirion detector did not measure any neutron emission.

The MICROSPEC2 BTI detector — positioned alongside the HDS Mirion detector, as indicated in the figure, in quadrant II at a distance of 30 cm from the sonicated AISI 304 steel bar — recorded a neutron pulse count of 110 at the peak maximum, corresponding to an energy of approximately 0.7 MeV. The total dose measured by the MICROSPEC2 detector during the experiment was 0.024 microsieverts, while the background measurement, conducted over 15 minutes, indicated a dose of 0.003 microsieverts. The counts

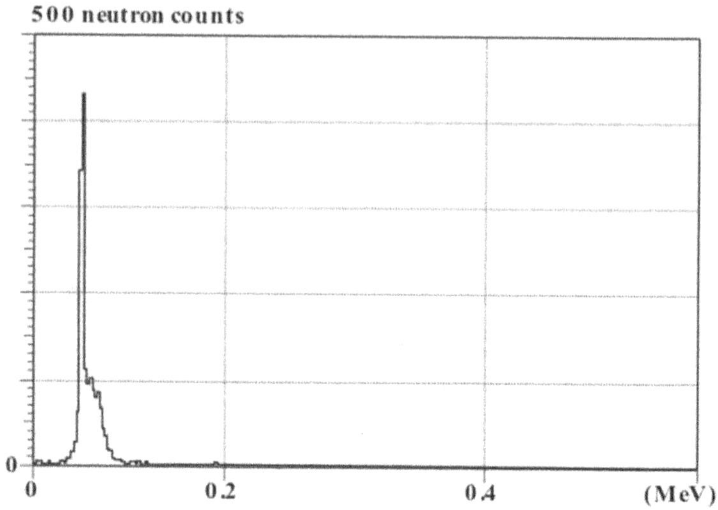

(a)

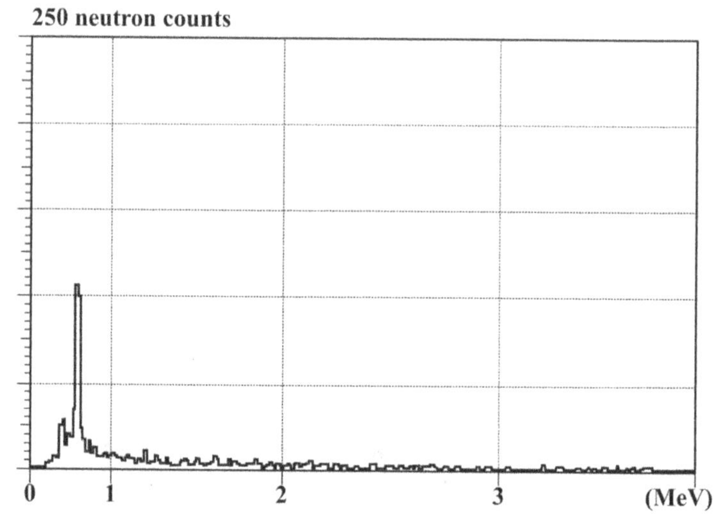

(b)

Fig. 6.11. Neutron counts are shown on the y-axis, and energy (MeV) on the x-axis. (a) He-3 gas detector (0–0.5 MeV). (b) NaI(Tl) detector for liquid scintillation in the range (0–4 MeV) low counts continue up to 16 MeV.

from the CR-39 AB detectors arranged around the steel bar were divided, as mentioned earlier, into quadrants. These comprised four 90° quadrants, each containing four CR-39 AB detectors, oriented South-North and proceeding counterclockwise.

Table 6.1 below summarizes the obtained data, including dose and dose rate measurements.

For the entire 360° plane, the measured dose was 49 microsieverts, corresponding to 981 microsieverts/hour, with an integrated neutron flux of $5.8 \times 10^6\,\text{n/cm}^2$, a neutron count rate of 294 n/sec, and an average neutron energy of 2.8 MeV. Subsequently, a further analysis of the obtained data was performed, focusing on the results of each of the 16 CR-39 AB detectors, numbered from 1 to 16 (see Fig. 6.12). The analysis was conducted in a counterclockwise direction, starting South-North.

The results are presented in Fig. 6.13, where each detector's data are highlighted, including the progressive detector number, the measured dose in microsieverts and the angle width measured relative to the conventional South-North reference axis. The length of the solid lines radiating from the center in a radial direction represents the measured dose analogically. This dose calculation includes the fraction of neutrons absorbed by the 3 cm thick Teflon sleeve. When analyzed in 22.5° intervals, both the asymmetric distribution of neutron pulses [7] and their orientation are clearly visible. The neutron emission reaches its maximum within a 90° sector between 157.5° and 247.5°, with a total dose of 110 microsieverts.

Furthermore, as shown in Fig. 6.13, it is evident that two nearly orthogonal directions exhibit specific characteristics: in one direction, numbered as 6 and 14, the dose is nearly zero, in the orthogonal direction, numbered 2 and 10, the dose is at its maximum.

Table 6.1. Summary of the data from the analysis of the CR 39AB, passive neutron detectors, as per Fig. (6.12)

	Unit	I Quadrant	II Quadrant	III Quadrant	IV Quadrant
Dose	μSv	12.6	13.0	21.0	2.4
Dose rate	μSv/h	252.3	259.0	421.3	48.0
Fluence n	n/cm^2	20.0×10^5	20.8×10^5	33.3×10^5	3.0×10^5
Counts	(n/sec)	78	76	126	14
Energy n	MeV	$< 1.0 >$	$< 1.7 >$	$< 2.8 >$	$< 0.3 >$

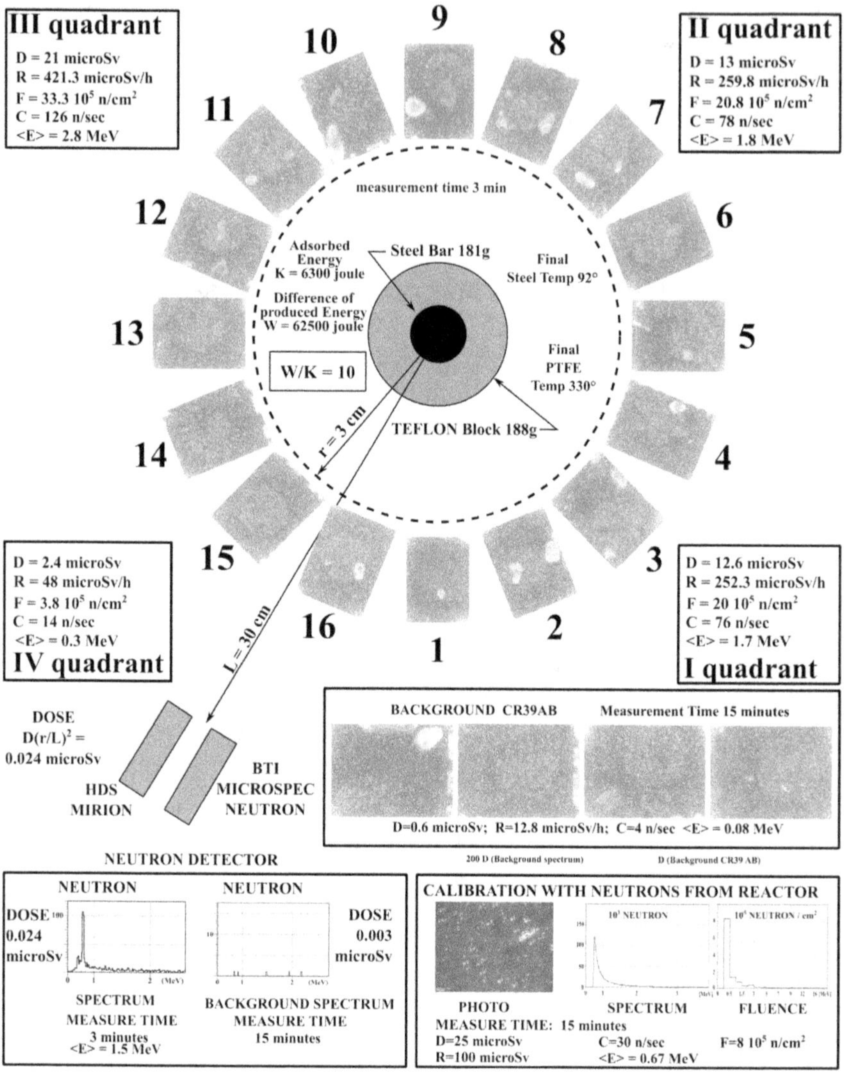

Fig. 6.12. Horizontal view of the sonicated rod system inserted into the PTFE cylinder — detectors. Outside, counterclockwise, in the South — North direction, from 1 to 16 the images of the CR-39 AB with the tracks released by the neutron impact.

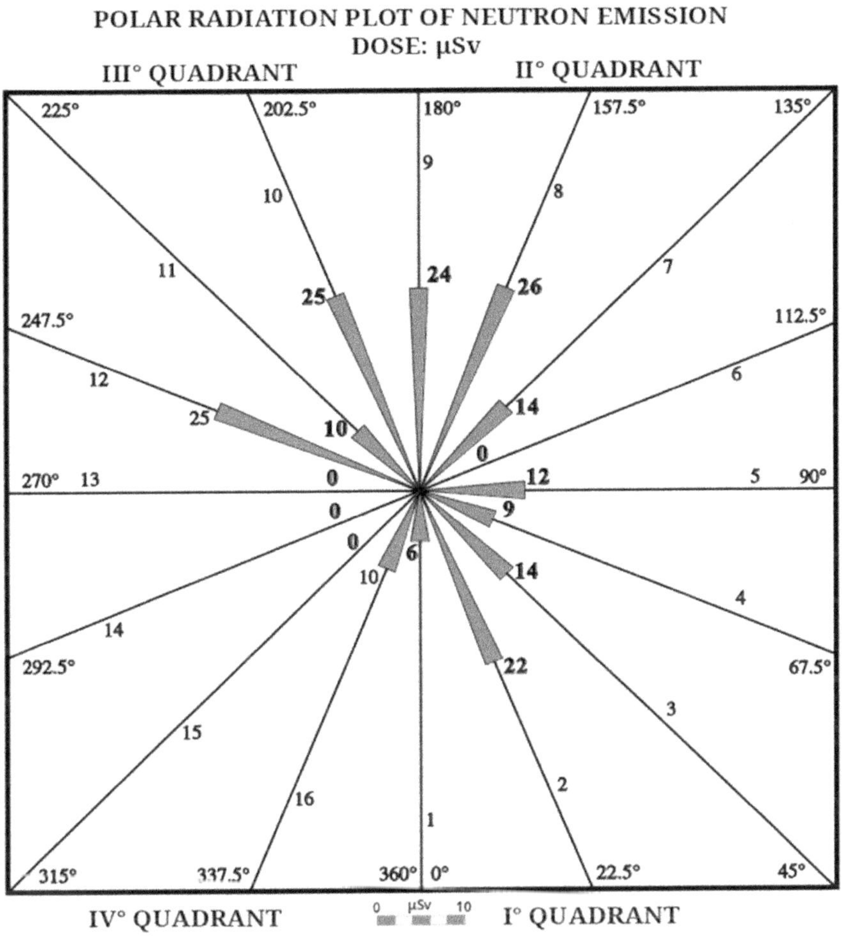

Fig. 6.13. Spatial Asymmetry of Neutron Burts from DST Emissions, Polar diagram of the Asymmetric neutron emissions from sonicated steel.

This demonstrates that the neutron dose is distinctly anisotropic and even asymmetric.

From all the above observations, it seems reasonable to conclude that the asymmetry and anisotropy of the measured neutron emission, along with the consistent absence of gamma radioactivity, suggest that this phenomenon does not involve nuclear fission or fusion reactions occurring in the leptonic energy domain, as described by the classical or standard model.

Instead, these observations point to reactions occurring in the hadronic energy domain for metric energies >367.5 GeV, in accordance with the Deformed Space Time (DST) theory. For the all above reasons we speak of DST reaction producing DST emission of neutrons which are 'new' kind of neutron emissions. About the DST reactions we can say that they can produce not only neutrons but other particles too. The DST reactions can be divided into two main kinds : the 'nucleosynthesis' when two or more nuclei undergo a union giving rise to nuclei of higher mass than the parent nuclei and the 'nucleolysis' when parent nuclei split in nuclei having lower mass than the parent ones. In both cases nuclear particles and fragments can be resulting product which do not leave the resulting nuclei neither in excited state nor in a metastable state and therefore without any gamma rays emission.

6.3 Detectors Description

Due to novelty in using thermodynamic neutron detectors, the so called Defenders, and the employ of 'photographic' neutron detection by using the Polyallil Diglycol Carbonate (PADC) plates CR-39 AB, we describe their main features in order to make the reader more acquainted with the detection techniques needed to face the task of measuring a 'new' kind of neutron emission coming in 'bursts'. These neutron bursts give rise to a new kind of neutron emissions which has very peculiar properties. Lack of synchrony, since the emission begins well after the starting of DST thrust, a sort of inertial time delay in the emission, then the emission lasts well after the end of the thrust, a sort of time latency of the emission. Moreover the time duration of the neutron burst can be below the time resolution of the detector and its electronics. Lack of spatial isotropy and presence of spatial asymmetry, which produced blind spots for the detectors positions where the signal is absent. At last the energy distribution of the neutrons in the burst is polyenergetics and shows a lack of regularity and homogeneity. Summarizing a true nightmare of the detection. At the end we do some remarks about the inner structure of the neutron bursts from the point of view of the energy. Looking in this order, growing energy, to Figs. 6.11(a), 6.11(b), 6.10(a), 6.10(b), we cannot discriminate neither if a single burst contains all these neutrons in different energy ranges nor if there are different bursts each made up of neutrons having only one energy range.

The Defender and Defender XL detectors (larger in size) are passive neutron detectors produced by BTI (Bubble Technology Industries). They consist of minute droplets of a special (superheated) liquid uniformly dispersed in an elastic polymer gel. When neutrons strike the droplets, they form small gas bubbles that remain fixed, trapped within the polymer. The number of bubbles is proportional to the number and energy of the neutrons, allowing a direct correlation between the bubbles and the neutron parameters. Each individual detector is provided with its own specific calibration: bubbles per milliRem and bubbles per microSv. This makes it possible to convert the bubble count into an equivalent dose. The Defender detectors are sensitive to neutrons in the energy range of 10 keV to 15 MeV and are immune to the effects of gamma radiation.

CR-39 PADC detectors are passive plastic detectors made from a polymer: $C_{12}H_{18}O_7$, with a density of 1.3 g/cm^3. They are used for the detection of heavy charged particles, for which they have high sensitivity. Charged particles are directly detected, while neutrons are detected via a secondary reaction caused by Boric Acid (H_3BO_3) applied as a hot coating on the surface of the polymer.

Upon impact, neutrons induce the nuclear reaction $B^{10}(n,\alpha)\ Li^7$ in natural Boric Acid (19.9% of B^{10}, 80.1% of B^{11}), where the emitted alpha particles leave tracks in the CR-39 plate. The tracks left by the charged particles become visible after chemical treatment, which is carried out by immersing the detectors in a NaOH (Sodium Hydroxide) solution for 4 hours in a thermostated bath at a temperature of 82°C. After the described treatment, the detectors are dried, observed under a microscope, and photographed. The energy range to which CR-39 detectors are sensitive is very broad, ranging from several tens of keV (40 keV) to several MeV. In our case, a variant of the CR-39 PADC, called CR-39 AB, was developed using boric acid powder surrounding it. Moreover an other model was tested where the polymer $C_{12}H_{18}O_7$ is replaced by a standard photographic film, consisting of an acetate substrate coated with a layer of AgCl (Silver Chloride). The photographic film is surrounded by finely powdered Boric Acid. As mentioned earlier, neutrons induce alpha particles that expose the photographic film, which is then developed, observed under a microscope, and photographed.

References

[1] F. Cardone *et al. Phys. Lett. A*, 373, 862–866, 2009.
[2] G. Cherubini *et al. Annales de la Fondation Louis de Broglie*, 34, 138, 2009.
[3] F. Cardone *et al. Mod. Phys. Lett. A*, 27, 18, 1250102, 2012.
[4] F. Ridolfi *et al. J. Adv. Phys.*, 2, 40–44, 2013.
[5] A. Rosada *et al. Detection*, 1, 30–35, 2013.
[6] A. Rosada *et al. J. Adv. Phys.*, 7, 1–9, 2018.
[7] E. Santoro *et al. Mod. Phys. Lett. B*, 29, 14, 1550067, 2015.
[8] F. Cardone and A. Rosada. *Mod. Phys. Lett. B*, 30, 28, 160346, 2016. DOI: 10.1142/S0217984916503462.
[9] G. Cherubini and A. Rosada. *Proceedings of 10th Symposium Vigier* p. 60–65. *Unified Field Mechanics II,* Ancona (Italy), 2016.

Chapter 7

Alpha Particle Emissions

7.1 Alpha Impulses in the Time and Space

We have seen in Chapter 6 the emission of neutron bursts from solutions of water enriched by iron salts and from basalt rods after a treatment by ultrasound [1]. Cavitation, which brings about a violent collapse of gas inside the material under ultrasound, was assumed to be responsible of that effect. Besides, similar evidences of neutron bursts were obtained by fracturing granite samples and we will see in the next Chapter 8. The application of ultrasound to bars of iron or steel resulted in the emission of neutron bursts as well. The spectrum of this peculiar emission was measured and the appearance of circular, macroscopic and regular damage spots was also observed on the bar surface. The superficial elemental microanalysis showed that the concentrations of chemical elements in these spots are different from those found in the undamaged areas. Therefore along with the emission of neutrons, some nuclear transformations occurred as well.

It has been performed an experimental investigation aimed at testing if loading-deloading cycles, slower than those of ultrasound, applied to steel bars can induce emission of particles others but neutrons. The aim was to slowdown the rate of the nuclear reactions with respect to the case of ultrasonic pressure in order to get emissions without the severe 'burst' structure, detected during the aforementioned neutron experiments, and therefore more easily detectable by standard instruments. The starting point was to consider two following reactions indicated as possible Deformed Space Time (DST) reactions that may take place and account for the emission of neutrons and for the microscopical decrease of iron, see Eqs. (6.2) and (6.3). In the second reaction, alpha particles appear as reaction product.

$$^{56}Fe^{30}_{26} \to 3^{12}C^{6}_{6} + ^{16}O^{8}_{8} + 4^{1}n^{1}_{0}$$

$$^{56}Fe^{30}_{26} \to 2^{12}C^{6}_{6} + ^{16}O^{8}_{8} + 3^{4}\alpha^{2}_{2} + 4^{1}n^{1}_{0}$$

The tests were based on the parallelism between bubble collapse in liquids and gas porosity collapse in solids, due to the porosity they present after the production process. Two tests have been set. The first studied the ionizing particle emission and their anisotropy. The second aimed to a systematic study of alpha emissions and the role played by the pressure in catalizing DST emissions [2].

Different cylindrical bars 2 cm in diameter and 20 cm long made by AISI 304 steel have been submitted to cyclic compression stress by using a press with its connected control system and data recording software. The results obtained by these two bars, labeled as A and B, are reported in Fig. 7.1. During the test each bar sample was compressed to a maximum load of 130 kN, lower than the corresponding yield load evaluated to be about 160 kN, with the same sequence of

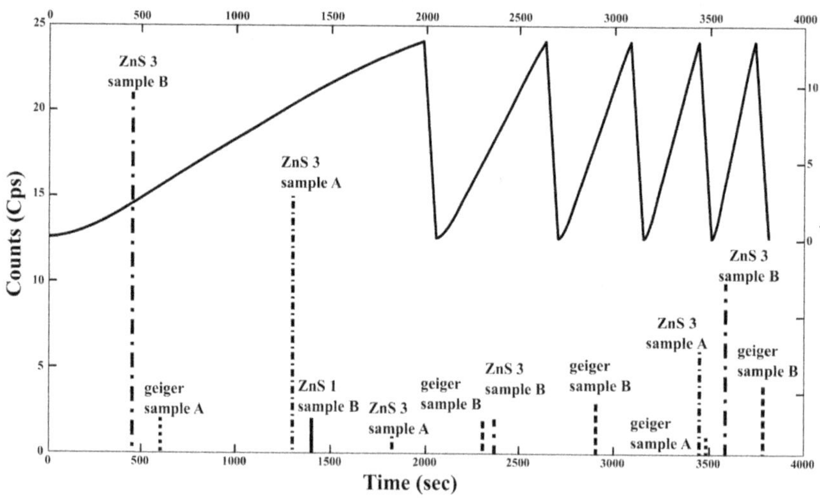

Fig. 7.1. Spikes of cps (left scale) after background subtraction and corresponding loading deloading cycles (right scale) as functions of time. The two steel samples are indicated as A, B. The three ZnS detectors are indicated as 1, 2 and 3.

cycles as shown in Fig. 7.1, changing the piston velocity five times from 1 micron/sec to 5 microns/sec, with steps of 1 micron/sec. Before each sequence, a load of 300 N was applied on the sample within a time interval of 60 seconds.

This pre-compression, or preload, also accounts for preloading a certain amount of energy density in the bar according to the prescriptions of the Deformed Space Time theory of DST reactions, as already done for the pneumatic contact between the sonotrode and the steel bar in the neutron experiments, see Fig. 6.7. Three alpha particle detectors, labelled as 1, 2 and 3 and a Geiger detector all measuring the count rate in counts/sec (cps) were placed as close as possible to the bar surface, less than 1 cm the distance from the detector active surface. The position geometry of the three alpha detectors was such to cover a 360° polar view, following what done for the neutron experiments, see Fig. 6.12. The measurements (cps) were recorded as functions of time. The alpha detectors were ZnS(Ag) scintillators coupled with a multichannel counter system as signal amplifier. The dark current, the thermal noise of photocathodes, of ZnS(Ag) was of 0.05 cps as specified by the manufacturer and was negligible with respect to the measured background. The Geiger counter was implemented with the related detector for Alpha, Beta and Gamma radiation. At the end of all the loading cycles, the outputs from the scintillators and the Geiger counter, which were always kept in the same operating position, were also recorded using a sample of Teflon placed in the position of the steel bars and was cyclically loaded. These last measurements lasted 3900 seconds, like each of the measurements with the steel bars, with the same five strain rates. In this way the experimental background was so recorded for each detector when the loading system was operating. This procedure was aimed at taking into account and eliminating the contributions from the instrumental backgrounds, the laboratory background and also from possible ionizing particles coming from the press during operation. Figure 7.1 reports the spikes obtained for the studied steel samples after the background subtractions together with their positions in the frame of the five loading/unloading cycles, with loading strain rate of $1\,\mu$ m/sec, $2\,\mu$ m/sec, $3\,\mu$ m/sec, $4\,\mu$ m/sec and $5\,\mu$ m/sec, respectively. The spikes of bar sample A in the first cycle loading rate, $1\,\mu$ m/sec, are reported in Fig. 7.2 together with the corresponding displacement/load hysteresis. In the first test the non-zero signals,

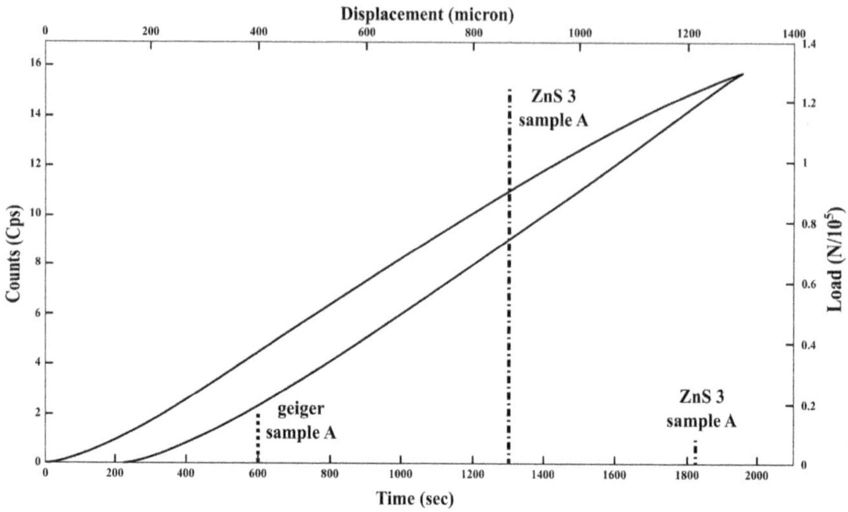

Fig. 7.2. Displacement–load curve (upper and right scales) of the first cycle of sample A and corresponding spikes (left scale). The peaks have been recorded during the loading phase. The time (lower scale) refers to the loading phase.

obtained after the background subtraction, have to be considered as an evidence of ionizing particles emitted from the steel bars subjected to compressive stress and strain varying in time. The alpha emission occurs, due to the positive response of alpha scintillators and also because the counts from Geiger are of the same order of magnitude as those from ZnS counters, thus suggesting that the same particles are detected. Therefore in the second test CR39 polycarbonate alpha detectors have been used as a cross-check of alpha emissions.

The emissions recorded during the first test sets of the whole experiment still show a spatial anisotropy, in analogy with what occurs for neutron emissions, as it is well evident from Fig. 7.1 where the majority of the spikes are from scintillator detector number 3. In fact, as we see in Fig. 7.1, looking at the emissions recorded for each sample, the anisotropy of detection suggest the possible anisotropy of the emissions. Moreover a noticeable fact, which clues the correctness of the theory of the DST reactions, is that the alpha emissions occurred systematically at the middle of compression (loading) or decompression (release) as it is well evident again from Fig. 7.1. This fact means that the pressure or better the energy density gradient plays a pivotal role in triggering these nuclear emissions, according to the theory of the Deformed Space Time for the nuclear interaction. In particular the loading and displacement of bars versus time as in Fig. 7.2, can suggest that the total emissions occur after the overcoming of energy threshold leading to the Local Lorentz invariance (LLI) violation that drives the occurrence of the DST reactions.

In the second test four AISI304 steel bars of cylindrical shape with 2 cm in diameter, 20 cm long and about 0.5 kg in weight were submitted to cyclic compression stress by the same press endowed with the same control and recording systems. Two of the bars, hereafter labelled A and B, were first pre-compressed for 1 min at 300 N and then subjected to five consecutive fatigue cycles. Each cycle consists of a compression phase from 300 N to 160 kN and a fast stress release phase, about 10 s, from 160 kN to 300 N. A different strain rate was applied in each of the five loading stages: 1, 2, 3, 4 and 5 μm/s, respectively. After these treatments the two bars were visibly deformed. In order to avoid such a macroscopic deformation, two other bars, henceforth labelled C and D, identical to the previous two, were compressed between 300 N and 130 kN, instead of 160 kN, following exactly the same sequence of cycles. In all cases, the strain and stress were automatically recorded as functions of time.

The experimental incertitude of the load was between 0.5% and 1% and the resolution of the displacement was 0.1 μm. During the fatigue cycles, were monitored the possible emissions of ionizing radiation from the bars by three alpha particle counters having the same geometrical arrangement as in the first test and equally named detectors 1, 2 and 3, and a Geiger counter. All of them were placed almost in contact with the bars and their amplified outputs were recorded as functions of time.

In the case of the fatigue tests with the samples A and B, the Geiger counter was not used. The three alpha particle counters were ZnS(Ag) scintillators of the same model. Their dark current, the thermal noise of photocathodes, as specified by the manufacturer, is 0.05 cps which is a negligible value with respect to the measured background. In order to establish the experimental zero of each detector when the press was working, it has been again used a Teflon sample which was subjected to the same treatment of the bars C and D. The knowledge of the experimental zero for each detector and of its statistical variations allowed again to take into account and hence to eliminate, by subtraction, the contributions from the instrumental and laboratory background and also the contributions from ionizing particles possibly coming from the press rather than from the iron samples.

In the second test besides these active detectors, a passive type of alpha detector was also used. It is commonly known as CR39 or Polyallil Diglycol Carbonate (PADC) detector. It is a plate of polycarbonate commonly used to detect alpha particles emitted from ^{222}Rn for environmental monitoring purposes. Its active surface is $1\,\text{cm} \times 1\,\text{cm}$ and its thickness is $1\,\text{mm}$.

Two CR39 plates were used for each of the four bar samples A, B, C, D. They were fixed tangent to the lateral surface of the cylindrical bar with two sides horizontal and the other two vertical. One was fixed to the upper part of the bar and one to the lower part. They were kept in these positions from the beginning to the end of the experimental sequences: preload, five cycles and final holding at 300 N. Other two CR39 plates were used to record the background during the fatigue tests on the four samples and the Teflon one. In particular one CR39, placed horizontally and far from the working press, recorded the background during the treatment of the samples A and B and a second CR39, placed with the same orientation and at the same distance from the press, recorded the background during the Teflon and sample C and D treatments. The count-rates of alpha particles, measured for the four samples, see Fig. 7.3, were obtained by subtracting the highest count-rate value of all the experimental zeroes, obtained by fatigue test of a Teflon sample, from each count-rate value collected during the fatigue tests on each of the four bars. Only the positive count-rate values surviving this subtraction can be considered as possible evidences of ionizing particles emitted from the steel bars. As usual this procedure applied to the data, performs a very severe cut because it completely disregards the different efficiency of the three ZnS(Ag) counters and establishes as experimental zero the highest possible, or rather the most pessimistic, background value.

The same procedure of background subtraction, applied to the count-rate values of the ZnS(Ag) scintillators, was also adopted in the case of the Geiger counter. Of course, the background level, the experimental zero, of the Geiger counter was different from that of the scintillators, because of the differences in the operating mode, the accepted energy range and the type of ionizing particles they

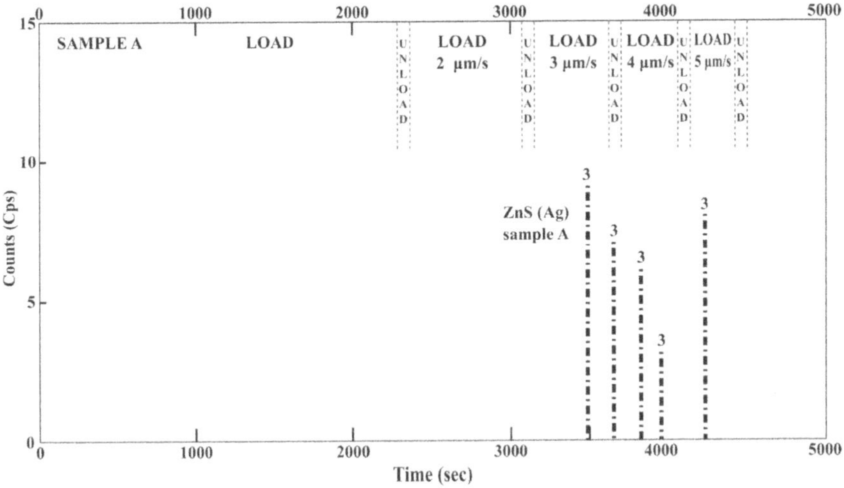

(a)

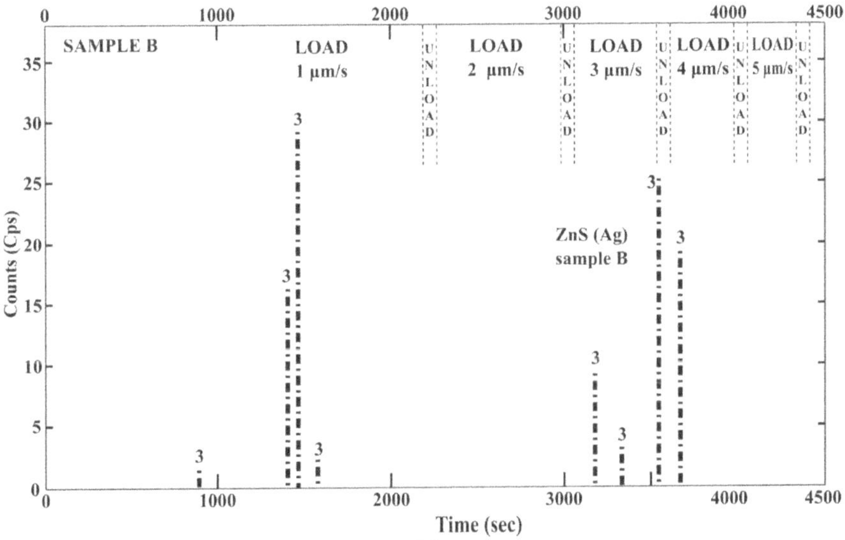

(b)

Fig. 7.3. (a) Sample bar A, counts of alpha detector ZnS(Ag), the number 3 refers to 3rd scintillator; (b) Sample bar B, counts of alpha detector ZnS (Ag) the number 3 refers to 3rd scintillator; (c) Sample bar C, counts of alpha detector ZnS(Ag) the number 1 and 3 refer to 1st and 3rd scintillator; (d) Sample bar D, counts of alpha detector ZnS(Ag) the number 3 refers to 3rd scintillator; (e) Sample bar C, Geiger intensity peaks; (f) Sample bar D Geiger intensity peak.

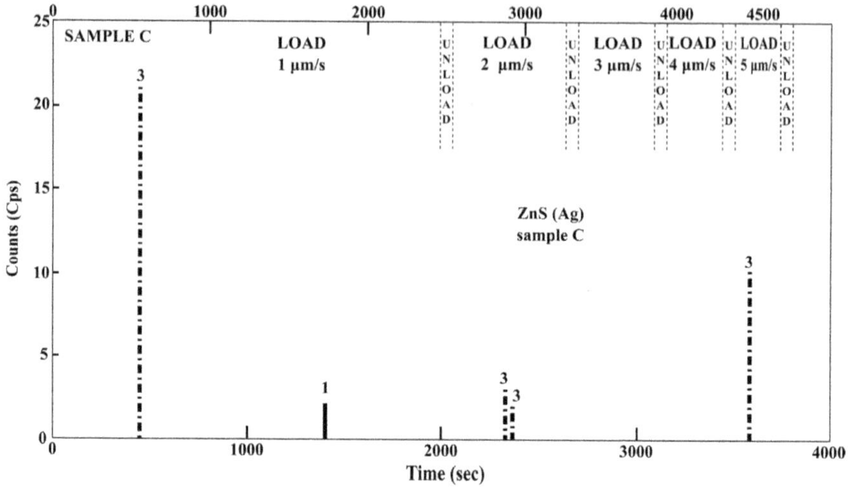

(c)

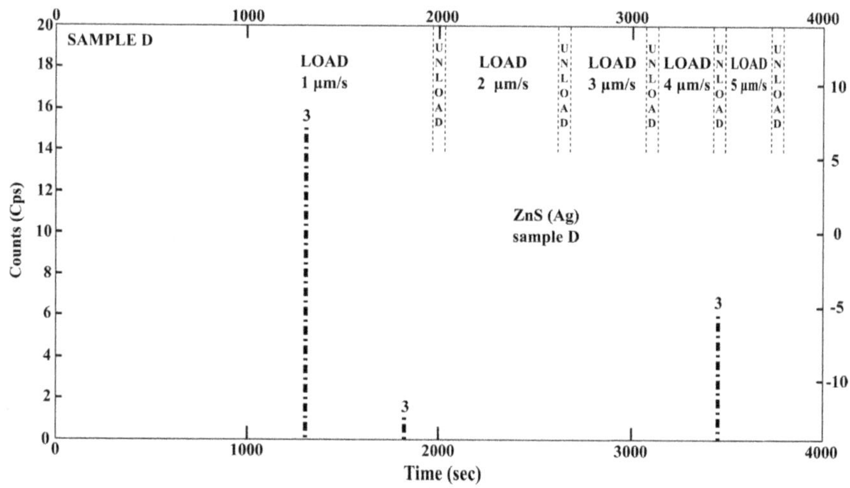

(d)

Fig. 7.3. (*Continued*)

can detect. The obtained intensities from the investigated samples are reported in Fig. 7.3. After the background subtraction, the peak intensities of the Geiger signals are not higher than those obtained in the case of the ZnS(Ag) scintillators.

The CR39 plates were developed by using a standard procedure: they were immersed and kept for 4 h in 3 litres of a six molar solution of NaOH, at a temperature of 90°C. After being taken out of

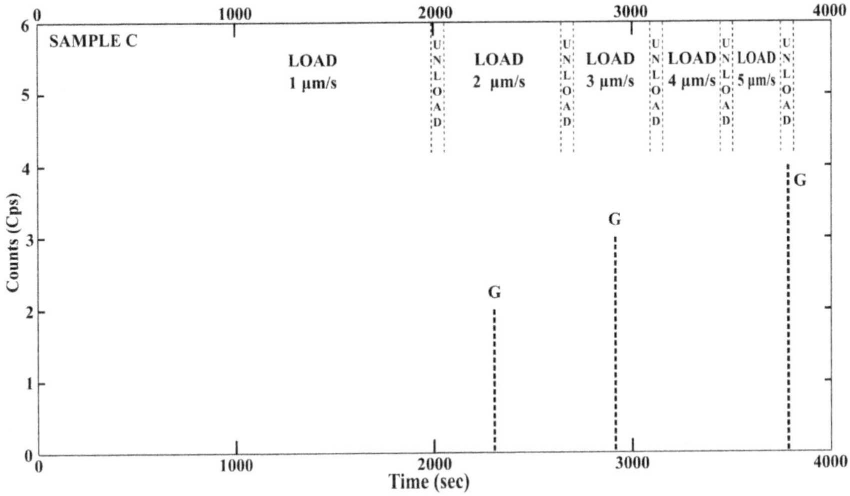

(e)

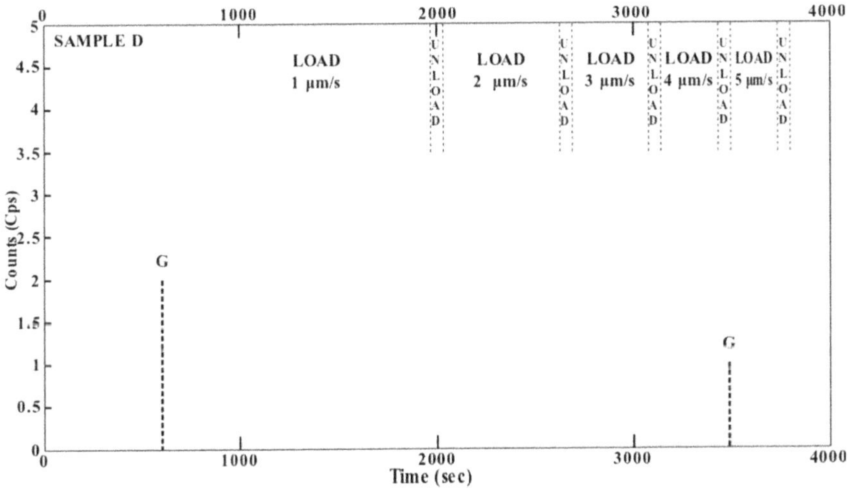

(f)

Fig. 7.3. (*Continued*)

the solution, the plates were rinsed in bidistilled deionized water and dried at room temperature (20°C). The plates were then observed in reflected light optical microscopy. For each of them, a camera was used to take two pictures: the first was obtained by focusing the photo camera on the upper surface and the second on the lower one of the plate. Changes in its molecular structure, either induced by ionizing particles or by other agents, create dark regions on the surface. These regions are selected and analyzed. The following criterion was used to select the spots to be considered as tracks of alpha particles emitted from the bars. As we have just said, for each CR39 plate two pictures have been taken, the upper and the lower surface. Only those dark points visible in both the cases, focus on upper and on lower surface, were taken into account, thus disregarding all those having an image only on one of the two sides. This strict method aimed at eliminating possible artefacts, that would lead to overestimate the number of alpha particles. Moreover, all the spots with a circular shape were not count, since they were considered as background signals, usually found in all the pictures.

Different instruments were used in this second test in order to check whether alpha particles would be emitted when appropriate pressure variations were exerted on steel bars. In particular, the three ZnS(Ag) scintillators are alpha active counters, the Geiger counter with its related electronics is sensible to ionizing radiation alpha, beta and gamma radiation, while the CR39 plates are passive detectors widely used for neutron and alpha detection. Concerning the images obtained from CR39 in close contact with the bars, some darker spots are visibly characterized by shapes different from those observed in the background. Therefore, it cannot be excluded that other kinds of ionizing particles, like say the protons, crossed the films and are responsible for such spots.

In a total time of $20{,}000 \pm 2000\,\text{s}$, 13 tracks were recorded in an area of $2.00 \pm 0.16\,\text{cm}^2$, the area of each plate is about $1.00 \pm 0.08\,\text{cm}^2$. The count rate of those tracks is thus about 4×10^{-4} track/$(\text{s} \cdot \text{cm}^2)$ ($\pm 30\%$), which is similar to the rate of bursts obtained in the case of the electronic ZnS(Ag) counters: 0.3×10^{-4} bursts/$(\text{s} \cdot \text{cm}^2)$. In fact, 21 bursts were recognized in $16{,}540 \pm 10\,\text{s}$ and the cross-section area of the sensible volume of each of the three counters was $50 \pm 5\,\text{cm}^2$ with a 30% efficiency, evaluated in the case of Pu(239) source. Neither the sole results of the CR39 detectors nor those of the ZnS(Ag) counters are a clear evidence that alpha particles are produced. However, those two different instruments, based on different detection procedures, both designed for alpha particle detection, give a positive response in the case of the four bars, if compared to the reference background. We consider this fact as a strong evidence of alpha particle emission.

A further support comes from the measurement of the Geiger counter. In fact, also in this case positive signals were obtained, see Figs. 7.3(e), 7.3(f). The particle counts registered by this detector were 14×10^{-4} bursts/$(\mathrm{s} \cdot \mathrm{cm}^2)$, five bursts in 7800 s on a 6.4 cm^2 active area, with an average efficiency of 7%: 2% for ^{14}C, 10% for ^{90}Sr ^{90}Y, 7% for ^{239}Pu and 7% for ^{99}Tc, which is not much higher than those obtained with the other two techniques. This fact strongly indicates that most of the signal was related to alpha rather than beta and gamma particles. In fact, being the Geiger counter sensitive to the last three forms of radiations, the difference between its data and the scintillators, or CR39s, would have been much higher if only beta and/or gamma particles had been emitted.

Thus, the three techniques together strongly suggest that alpha particles were emitted. This conclusion is further reinforced if, in analogy with the case of the CR39 plates, the efficiency is not taken into account. In fact, in our counts a burst or a fraction of a burst are both evaluated as 1 unit. In this case the burst/$(\mathrm{s} \cdot \mathrm{cm}^2)$ are: 4×10^{-4} for the CR39 detectors, 0.1×10^{-4} for the ZnS(Ag) counters and 1×10^{-4} for the Geiger counter. Figure 7.4(a) reports the three dimensional Cartesian frame of reference chosen for evaluating the paths of the alpha particles. The contact region between the surface of sample A and the CR39 plate is reported as a vertical white line in the central part of Fig. 7.4(b). In the same figure the three alpha tracks are labelled with numbers 1, 2 and 3. From the images reported in Fig. 7.4 also considering the thickness of the plates, a three-dimensional reconstruction of the tracks was performed even in all CR39 plates used. This reconstruction allowed us to figure out the direction of incidence on the CR39 plate of the alpha particles.

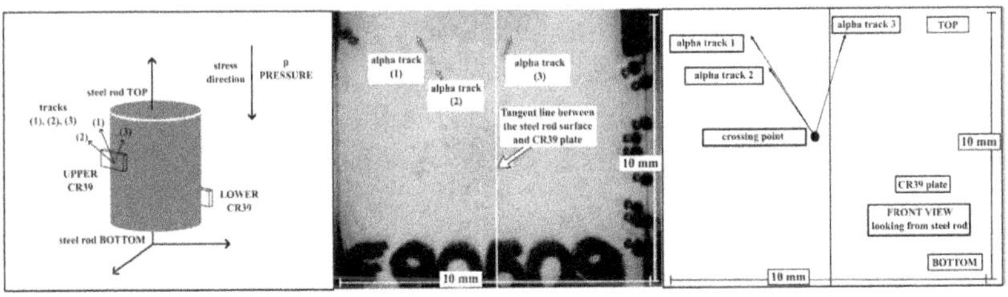

(a) (b) (c)

Fig. 7.4. Sketch of the chosen geometry for CR39 plates in the case of the sample "A" (a) and its relevance in the image of the upper position detector (b). Crossing point of the candidate tracks in a front view (c).

By prolonging these directions toward the surface of the bar, the front view is reported in Fig. 7.4(c), we noticed that the paths roughly met in a point belonging to the bar. This point could indicate the region where the deformed space time reactions, that produced the alpha particles, took place.

However, this hypothesis was rejected, because some millimetres of their paths would have been inside the steel, that would have stopped them unless one hypothesizes their energy to be enormously high. We rather think that the detected particles were generated close to the surface and had enough energy to cross at least 0.1–0.2 mm of air before hitting the plate. So, we estimate that their energy was higher than 40 KeV, corresponding to the minimum energy detected by the CR39 detectors. This conclusion is supported by applying a similar three dimensional-reconstruction procedure to the other CR39 plates referring to the other bar samples. It was not always possible to find a common crossing point, thus indicating that the emission came from different regions. A further reason is that all the paths are verified to start from the sample surface and, what is more important, they all came from a surface region close to the tangent line between the steel rod surface and the CR39 plate, see Fig. 7.4(b) and 7.4(c). These paths are crossing the thinnest layers of air and, as a consequence, the alpha particles had a relatively low amount of energy. The detected particles may also correspond to the low-energy tail or the high-energy tail of a larger energy distribution. However, the former hypothesis is less realistic, according to the just reported comments. Anyway, for all the tracks, it was possible to evaluate the angle θ between the strain direction and the alpha track direction. In the experimental set-up, the bottom part of the press was kept fixed, while moving the upper part. Thus the pressure front and the corresponding energy were moving from the top to the bottom, see Fig. 7.4(a), the values of the angle θ so evaluated range from 34° and 38° [3]. This evidence will be further considered in the next Chapter 8 aiming at investigating the asymmetry of the emission angle.

7.2 Energy Densities for the Emissions

We described in Section 7.1, and Chapter 6 a new kind of nuclear reactions, characterized by neutron and alpha emissions without concomitant gamma emission, these reactions were reported to occur after mechanical pressure or cavitation. After their experimental tests the occurrence of some fundamental structural and physical characteristics was suggested to be needed. They concern the energy density accumulated in the material, the number and dimensions of bubbles/cavities acting as microreactors [4]. In the following, we shall discuss about the size of the cavities, assumed of spherical shape, and the energy barrier to overcome. The first condition has been experimentally investigated in iron samples treated with ultrasonic waves and analysed by FESEM techniques [5]. Concerning the existence of an energy barrier, or better energy threshold, we take into account the Deformed Space Time theory, which gives a strictly relationship between the violation of the Local Lorentz Invariance (LLI) and the existence of energy thresholds, as we saw in Chapters 2 and 5. According to the results described in Section 2.5, a schematic sequence of the energy thresholds is reported in Fig. 7.5 together with the first evaluation of the corresponding energy densities. The 5th and 6th thresholds, and the non reported successive ones, are extrapolations according the recursive rules, see Fig. 2.2. We report some emissions from steel bars that are related to the third and fourth ones,

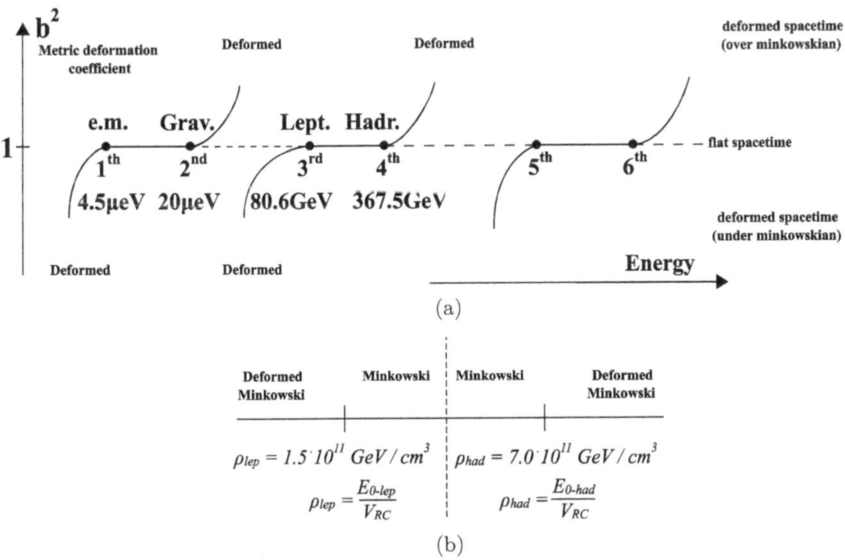

Fig. 7.5. (a) Energy values for the principal interactions; (b) Critical energy density for leptonic and hadronic interactions. V_{RC} is the Ridolfi Cavity volume, so named after F. Ridolfi [5].

corresponding to the leptonic and hadronic thresholds respectively. Nearby the third energy threshold, a leptonic type emission occurs, detection of alpha particles, while in the proximity of the fourth energy threshold, a hadronic type emission is expected bringing to neutrons detection. The related data analysis is aimed at evaluating the main physical parameters such as the energy density accumulated in the sample during the compression cycles. These physical quantities, obtained from the experiments performed on steel bars, are to be taken into the framework of the Deformed Space Time theory, in order to discriminate whether the emissions are due to leptonic or hadronic interactions in DST conditions [5]. The registered counts are then put in relation with the volume of the samples, the energy density accumulated at the emission and the elapsed time.

In order to describe the physics of the DST emission, we remind the phenomenon of cavitation, namely the collapse of gas within the cavities inside the materials induced by a pressure, wave pressure or mechanical pressure. As said in the phenomenology of DST emissions, these cavities can act like micro-reactors in which nuclear processes occur under suitable circumstances. The term 'micro' comes from their size, which corresponds to diameters of about 5–10 μm.

Let us recall in brief that the process of cavitation consists in subjecting gaseous liquids to elastic pressure waves of suitable power and frequency, in particular to ultrasound. The use of the term 'cavitation' is not fully correct, since it refers to a wide range of types of gas bubble collapses. We consider only the symmetrical spherical collapse, and in this meaning we use in this book the term 'cavitation'. Processes similar to those related to the 'cavitation' can also occur in solid materials when the gas inside the cavities of the bulk material collapses under the pressure waves, created by say a mechanical cyclic load. The collapse inside these cavities drives the reactions and then the emission of ionizing particles too. This cavitation, in fact, produces an extremely high concentration of energy per unit time in the collapsing bubble. The emission, however, only occurs if the energy density exceeds the critical energy density ρ_{int} corresponding to the $E_{0\,\mathrm{int}}$ threshold, see Fig. 7.5. Referring to our problem, as a first step we determine the energy density corresponding to the appropriate micro-cavities, called 'Ridolfi cavities' from now on, detected by FESEM in the iron sample subjected to ultrasound. In the case of leptonic interaction, the threshold energy is $E_{0\,\mathrm{lep}} = 80.6$ GeV $= 1.3 \cdot 10^{-8}$ Joule, the cavity volume is V $= 1.3 \cdot 10^{-16}\,\mathrm{m}^3$ and thus the leptonic energy density threshold is equal to $\rho_{\mathrm{lep}} = 10^8$ Joule/m^3. In the case of hadronic interaction, $E_{0\,\mathrm{had}} = 367.5$ GeV $= 5.88 \cdot 10^{-8}$ Joule and the corresponding energy density is $\rho_{\mathrm{had}} = 4.7 \cdot 10^8$ Joule/m^3. All these values corresponds to a pressure, we recall that, without loss of generality, the figure for the energy density also expresses a pressure according to the unit system used, therefore in MKSA the J/m^3 refers to N/m^2. These densities values, ρ_{lep} and ρ_{had}, are our starting point for the analysis of the data collected during the experimental tests of the cylindrical steel bars. As a second step, we consider what impulsive energy variation, corresponding to the work of an applied external force, induces the breakdown of Local Lorentz Invariance (LLI) and the departure from a flat Minkowskian metric.

As already illustrated all these energies are of crucial importance for the Deformed Space Time and the corresponding particle emission due to the DST emission. In fact, as seen the LLI violation needs one more dimension to be used as a physical variable describing the metric changes. The $E_{0\mathrm{int}}$ energy is related to this fifth dimension characteristic of the interaction metric. We are now able to describe the process of the DST reaction and to identify, on the basis of the energy density accumulated by the sample, which kind of interaction took place during the emissions of ionizing particles from the treated sample. We apply this theory to the results obtained from steel bars compressed by a cyclic load and from which emissions of alpha particles have been detected. In particular we will take into account the results reported in Figs. 7.5 and 7.6 and then we can determine the energy density accumulated in the sample at the time of the detected emissions in order to compare it with that calculated in correspondence of the interaction threshold.

Following the DST theory, we speculated whether the alpha emissions are due to leptonic or hadronic interactions in condition of deformed space time. To this aim, using the results shown in Fig. 7.5 and Fig. 7.6, we first calculated the accumulated energy density during every compression cycles. This is the ratio between the work performed by the compression load and the sample volume, varying all along the cycles. In Fig. 7.7 we report this energy density as a function of the time during the first loading phase of the D sample. The alpha emission peaks, obtained after the above-described subtraction, are also reported. The energy density at the time of the

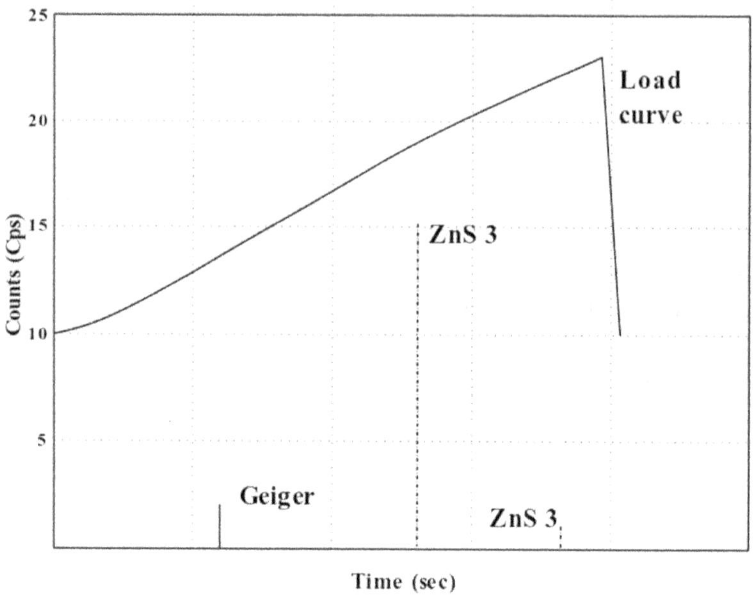

Fig. 7.6. Emissions recorded during the first cycle of the sample D by one of the scintillators and by the Geiger. The three events shown in this figure are those also indicated in Fig. 7.3(d).

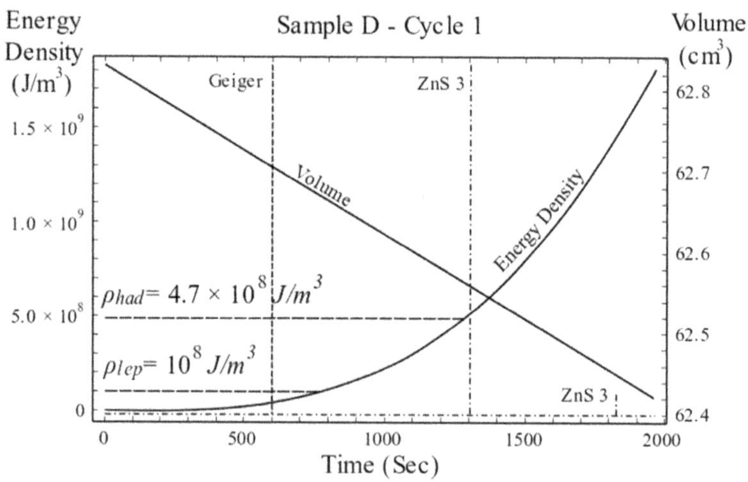

Fig. 7.7. Counts, energy density and sample volume during the first compression cycle of the D sample.

emission corresponds to the work of compression load integrated from the time zero to that moment divided by the volume, which also is reported in the figures, at that time. Thus the energy density and the volume corresponding to the emissions could be evaluated. The so obtained values of the energy density and of the volume at the moment of the emission provide the exact information about the five fundamental coordinates, say space, time and energy, necessary to describe the characteristics of the DST emission. The energy density at the emission is reported in Tables 7.1 and 7.2, for the sample D and C respectively. The number of emissions registered in the detector, indicated by N, and the corresponding time values are also reported. It is interesting to note that the most of the emissions correspond to energy densities in a range close to leptonic and to the hadronic thresholds $\rho_{\text{lep}} = 10^8$ J/m^3 and $\rho_{\text{had}} = 4.7 \cdot 10^8$ J/m^3 respectively. As an example, Fig. 7.7 shows the emissions detected by the scintillator ZnS number 3 and the Geiger counter during the first loading cycle of the sample D. The Geiger peak reported in Fig. 7.7 lies below the leptonic threshold, while the ZnS number 3 alpha detector peaks are situated one above the hadronic threshold and the other well above the hadronic threshold.

A detailed analysis of the region well over the hadronic critical density should be interesting, as neutron emissions could also be possible. To this aim, we note that a load close to the bar deformations was reached in these tests, thus other experiments reaching till to the bar rupture are needed. A large number of samples has to be considered, in order to obtain some information from the statistical distribution of data. The neutron emissions detected in these experiments

Table 7.1. Emissions data for Sample D.

Cycle 1	ZnS 3	Geiger
Emission number (N)	2	1
Emission time (Sec)	1303–1825	601
Energy density (J/m^3)	5.2 10^8–1.5 10^9	4.1 10^7
Cycle 1	ZnS 3	Geiger
Emission number (N)	1	1
Emission time (Sec)	316	342
Energy density (J/m^3)	3.1 10^8	3.3 10^8

Table 7.2. Emissions data for Sample C.

Cycle 1	ZnS 1	ZnS 3	Geiger
Emission number (N)	1	1	==
Emission time (Sec)	1405	455	==
Energy density (J/m^3)	6.3 10^8	1.4 10^7	==
Cycle 2	**ZnS 1**	**ZnS 3**	**Geiger**
Emission number (N)	==	2	1
Emission time (Sec)	==	276–312	246
Energy density (J/m^3)	==	6.4 10^7–9.2 10^7	3.3 10^8
Cycle 3	**ZnS 1**	**ZnS 3**	**Geiger**
Emission number (N)	==	==	1
Emission time (Sec)	==	==	209
Energy density (J/m^3)	==	==	6.3 10^7
Cycle 3	**ZnS 1**	**ZnS 3**	**Geiger**
Emission number (N)	==	1	1
Emission time (Sec)	==	75	274
Energy density (J/m^3)	==	8.6 10^6	2.8 10^8

obtained a meaningful link between the above reported conditions for alpha emissions and the neutron emissions from ultrasound sonicated samples and from fractured samples we also consider in the next Chapter 8 about the problem of emission asymmetry.

7.3 Statistics of the Emissions

We saw in the previous Paragraph that the different behaviour in the emissions process is related to the different energy densities characterizing the processes. These densities, in turn, are related to different energy thresholds, as they were obtained in the framework of the DST theory. Now we report the statistic results testing these phenomenological and theoretical predictions. According to DST theory, in fact, the different behaviour can be attributed to the existence of different energy thresholds for the deformation of the space-time metrics in the case of weak (leptonic) and strong (hadronic) nuclear interactions. Both below the leptonic threshold, and above the hadronic threshold, the emissions occur in DST conditions. This 'space' deformation justifies the emissions, which otherwise are not expected. In the case of cyclically compressed rods, the detection of alpha particles is related to energy densities below the leptonic threshold, and above the hadronic threshold. In those tests, loads close to the bar deformations were reached. Thus other experiments, reaching till to the bar rupture, were invoked in order to check if neutrons are emitted at higher energy density. In order to obtain some information from the statistical distribution of data, a large number of samples have to be considered. If neutron emissions are detected in these experiments, a meaningful link is obtained between the conditions for alpha emissions from compressed bars and the neutron emissions from ultrasound sonicated samples and from fractured samples [6]. Now we report the statistical behaviour of the DST emissions from steel rods when they are subjected to rupture. Measurements of ionizing and nonionizing particles were performed during the routine job of tension rupture tests at the Meccano SpA enterprise in Fabriano (Ancona, Italy) during 2013. In particular, 60 cylindrical steel rods were stretched to rupture. They had the same length, 100 cm, and three different diameters: 8, 10 and 12 mm. Three kinds of detector were used, a ZnS(Ag) alpha detector, a Geiger counter with related model of detector for alpha, beta and gamma radiation and an He^3 proportional neutron counter. A multi-channel counter system was used as signal amplifier for the three counters and also as power supplier for the alpha and the neutron counters. We are well aware that an He^3 gas detector for neutron is more sensitive to low energy range neutrons, see for example Fig. 6.11(a). Therefore the neutron measured in these tests are not the measure of all neutrons produced in all energy ranges during the rupture tests.

Figure 7.8 summarize the main results obtained from the ZnS alpha detector during the rupture tests of bars having diameter of 8 mm (left), 10 mm (center) and 12 mm (right). For each of the samples, the highest number of counts recorded in a second from the beginning to the end of the test was considered. This value is reported in the abscissa of the histograms, while the ordinate indicates the percentage of bars that produced the corresponding number of counts on the abscissa.

The results obtained with the Geiger counter are reported in Fig. 7.9. Those of the neutron detector are gathered in Fig. 7.10.

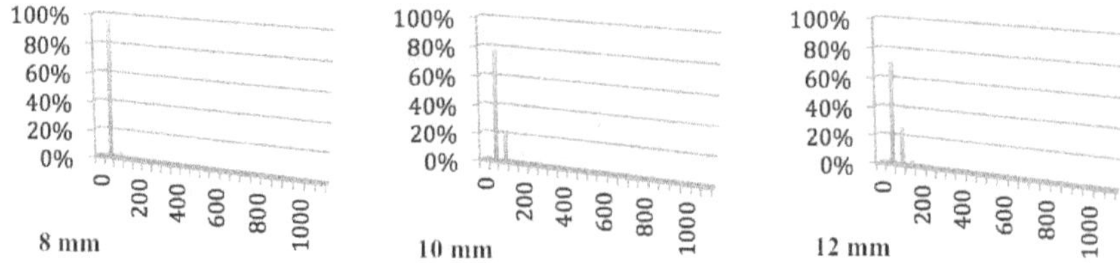

Fig. 7.8. Statistical information from the ZnS alpha detector. Percent of bars characterized by the maximum number of counts registered in a second reported in the abscissa. Steel rods having diameters of 8 mm (left), 10 mm (center) and 12 mm (right) are investigated. (The first interval, which is empty in these three cases, corresponds to no count.)

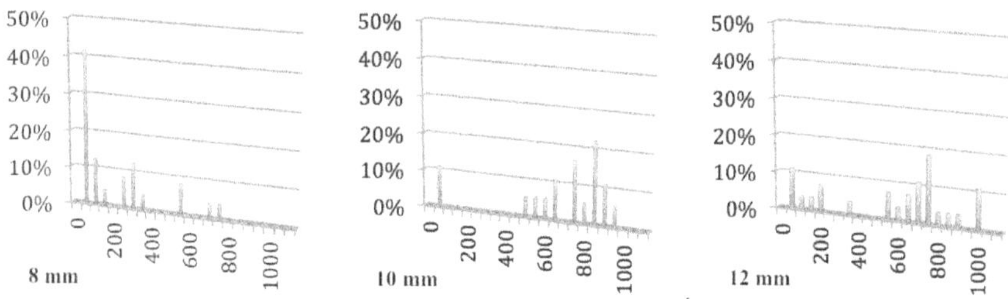

Fig. 7.9. Statistical information from the Geiger detector. Percent of bars characterized by the maximum number of counts registered in a second reported in the abscissa. Steel rods having diameters of 8 mm (left), 10 mm (center) and 12 mm (right) are investigated. (The first interval, which is empty in these three cases, corresponds to no count.)

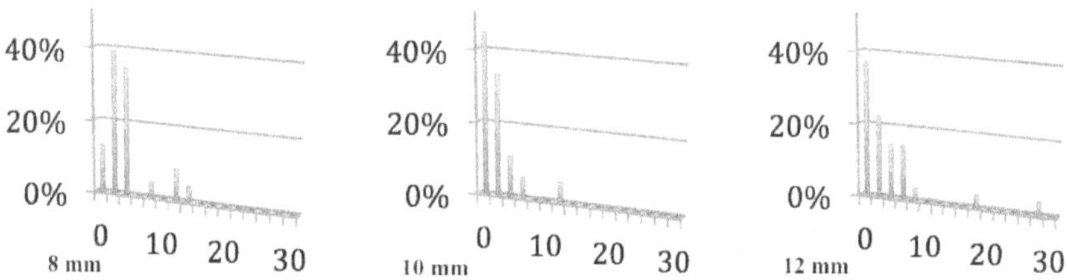

Fig. 7.10. Statistical information from the He^3 neutron detector. Percent of bars characterized by the maximum number of counts registered in a second reported in the abscissa. Steel rods having diameters of 8 mm (left), 10 mm (center) and 12 mm (right) are investigated. (The first column of each histogram corresponds to no count.)

The environmental background radiation was recorded at the beginning of all the sessions of measurements and after each rupture test. In this way it has obtained reference signals without contribution from the fractured samples. However, the distributions of the signals obtained after the rupture are similar to those reported in Figs. 7.8–7.10. We take into account the results of Chapter 6 where neutron emissions were induced by ultrasound after an 'inertial time' and continued also after switching off the ultrasound source for a 'latency time', also we consider that in this case a large number of tested samples were present in the laboratory and that the time between two tests was short. Therefore it can be assumed that the data obtained soon after the fracture are to be considered as measurements of the emissions caused by the rupture of the samples. This assumption is reinforced by the measurements performed at the beginning of the sessions, when no sample was yet tested: they indicate lower emissions and are in agreement with the background measured in the past in the same laboratory. Previous tests on cylindrical steel rods subjected to stress/strain cycles without rupture reported that the count rate of ZnS alpha detector and Geiger counter were of the same order of magnitude, some tents per second. From this evidence, it was deduced that the particles detected by the two counters are probably the same and in particular, after comparison with the data of CR39 films, it was inferred that they are alpha particles. In fact the number of alpha particles per second in unit area evaluated from the CR39 detectors is of the same order of magnitude of the values obtained from the two counters. By following this conclusion, we may assume that the columns in Fig. 7.9 corresponding to some tents of count/sec registered by the Geiger counter, are mainly due to the same kind of particles detected by the ZnS alpha counter and reported in Fig. 7.8. Beside that distribution at low counts, however, another distribution is visible in Fig. 7.9. It reaches some hundreds of count/s and its maximum value increases with the rod diameter. Such a bi-modal distribution is particularly evident in the central histogram of Fig. 7.9, corresponding to the 10 mm diameters. The higher count rates registered by the Geiger are to be attributed to its sensibility to alpha, beta and gamma radiation, while ZnS counter is mainly sensible to the alpha radiation. Thus it is possible to conclude that the histograms of Fig. 7.9 are to be attributed to two populations. One at a lower count rate, that was also observed in cycled samples not ruptured. The other, at a higher count rate, that corresponds to particles related to the event of the rupture. In particular, it is possible to suppose that they are electrons since the literature reports fracture induced electrons that are detected also after fracture, see [7]. A similar conclusion can also be drawn from the data of Fig. 7.10. Beside a large number of samples belonging to the population with a maximum of one or two counts per second, the first column of the three histograms of Fig. 7.10 corresponds to zero counts, the second to one or two counts, some cases are observed belonging to a population reaching count rates of some tents per second. This last result can be considered as an evidence for the emission of neutrons during the fracture/rupture phenomenon. A different behaviour between broken and not broken samples is well consistent with what described in Section 7.2 about energy densities. In fact, the different energy densities accumulated inside the samples in the two cases were suggested to correspond to different thresholds for the deformation of the space-time metrics. The statistical information, obtained on the features of nuclear particles emissions during rupture of steel bars, allows to make a comparison with the lower energy case of steel bars cyclically stressed without reaching the rupture point. This comparison is needed as a check for the theoretical prediction of neutron emissions at the rupture, in addition to the alpha emissions during the stress cycles as the latter correspond to lower density of energy. After analysing the signal distributions of a Geiger and a ZnS alpha counter, two populations of particles are deduced to be detected. One population mainly corresponds to alpha particles, as previously detected in stress/strain cycled samples that did not reach the rupture point. The other population is suggested to correspond to electrons emitted during and after the fracture. In the case of neutrons too, two populations are envisaged in the obtained distributions. One, corresponding to the larger number of samples, has a similar count rate as in not broken samples. The other, with a higher neutron intensity, is a hint for the emission of neutrons when the rupture conditions are reached. We remark that the count rates reported in the figures were obtained by detectors which are suitable for conventional continuous particle emissions. On the contrary, in the case of DST emissions due to DST reactions particles are expected in very short time pulses. The occurrence of neutron burst has been well examined in Chapter 6. Thus, as long as DST reactions are considered, the values reported in the abscissa in Figs. 7.8–7.10 are better to be considered as count rate of pulses made of many particles rather than particles per second.

References

[1] A. Manuello *et al.* Courtesy of private communication about Turin Polythecnic tests.
[2] G. Albertini *et al. J. Adv. Phys.*, 2(1), 20, 2013.
[3] G. Albertini *et al. Int. J. Modern Phys. B*, 27(23), p. 1350124, 2013. DOI: 10.1142/S021797921350124.
[4] F. Cardone *et al. Modern Phys. Lett. B*, 28(2), 1450012, 2014. DOI:10.1142/s0217984914500122.
[5] F. Ridolfi *et al. J. Adv. Phys.*, 2(1), 40–44, 2013.
[6] G. Albertini *et al. Modern Phys. Lett. B*, 28(5), 1450036, 2014. DOI: 10.1142/s0217984914500365.
[7] M. Osa and S. Shimamura. *Mater. Sci. Res. Int.*, 7(2), 111–115, 2001.

Chapter 8

Alpha and Neutron Angle

In the most general context of Deformed Space Time (DST) reactions [1], both for alpha particles and neutrons, we want to show that the violation of Lorentz invariance at the base of these emission processes is intimately connected to an angle of asymmetry of emissions.

In Chapter 7 we reported the analysis of the Ancona experiment regarding DST emissions of alpha particles and their direction of emission from compressed steel bars. In the present chapter we compare these results with those ones of the Cagliari experiment about DST neutrons emission during the failure of suitable material undergoing compression to rupture. We want to give a suggestion on the fundamental role that the angle of emission has in these processes that occur when the Lorentz invariance is violated. For the sake of brevity and simplicity the experiments on emissions of alpha and neutrons are named after the places where they have been performed.

Recovering the two main angles identified for the anisotropic emissions of the particles, alpha particles for Ancona and neutrons for Cagliari, we define a range of values within which to place the angle corresponding to the direction along which there is violation of Lorentz local invariance. As we will see in the next Chapters 13–15 the angles which occurred are connected with different electromagnetic experiments. For brevity, we present an example for each group of experiments. These experiments are significant because they show the behaviour of the emissions detected in all the measures.

8.1 Ancona Experiment: Alpha Particle Emissions

The experiment performed at Ancona in 2011–2012 have reached the goals of discovering both the DST emissions of alpha particles from compressed steel bars, see Chapter 7, and the diagonal direction of the alpha particles emitted with respect to the pressure direction [2]. Figure 8.1 reports the three-dimensional Cartesian frame of reference chosen for evaluating the paths of the alpha particles. The contact region between the surface of steel bar, sample A, and the CR39 plate of Polyallil Dimethil Carbonate (PADC) is reported as a vertical line in the central part of Fig. 8.2. In the same figure, the three alpha tracks are labelled with numbers 1, 2 and 3.

A three dimensional reconstruction of the tracks allowed us to figure out the direction of incidence on the CR39 plate of the alpha particles. By prolonging these directions towards the surface of the

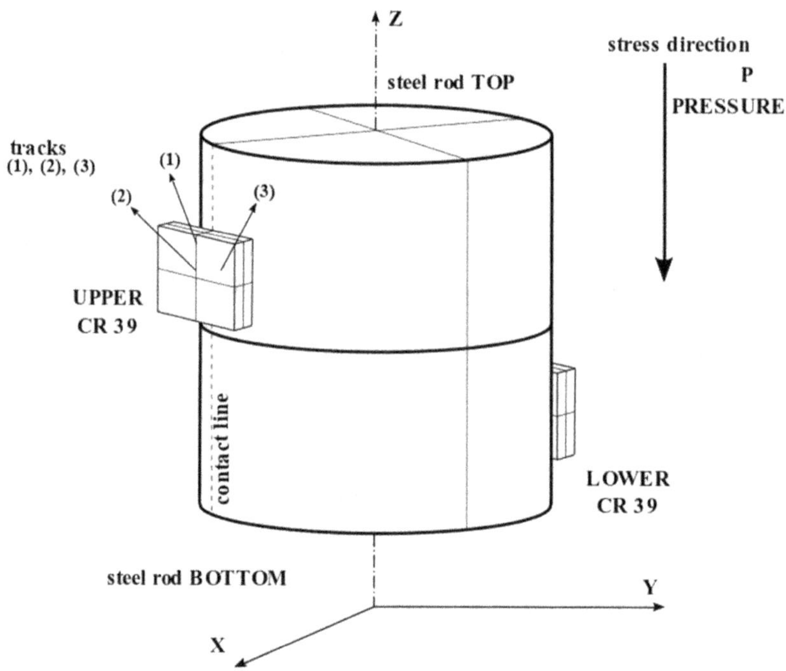

Fig. 8.1. The three dimensional Cartesian frame of reference chosen for evaluating the paths of the alpha particles, reported this the contact line between the CR 39 plate and steel bar surface.

bar as the front view is reported in Fig. 8.2, we noticed that the paths roughly met in a point belonging to the bar. This point could indicate the region where the DST reactions, that produced the alpha particles, took place [3].

However, this hypothesis was rejected because some millimetres of their paths would have been inside the steel, which would have stopped them unless one supposes their energy to be huge. Therefore it is well evident that the detected particles were generated close to the surface and got enough energy to cross at least 0.1–0.2 mm of air before hitting the plate. So it is possible to estimate that their energy was higher than 40 keV, corresponding to the minimum energy detected by the CR39 detectors.

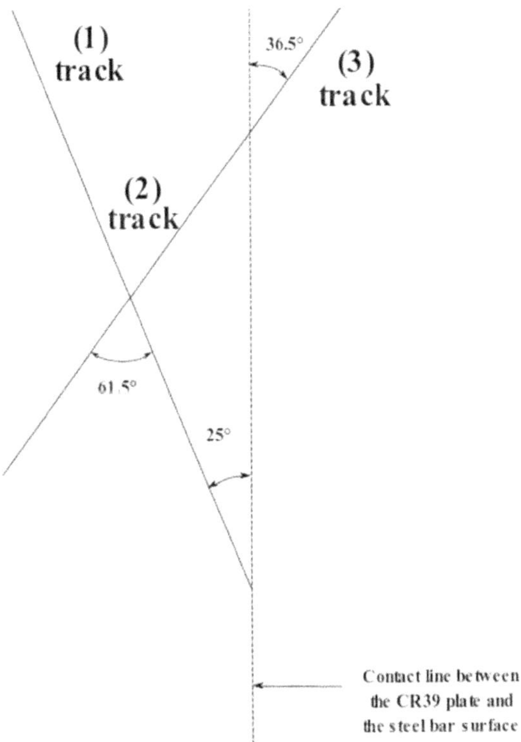

Fig. 8.2. Sketch of a CR 39 plate and the geometry of alpha tracks projected on the tangent plane referred to the contact line between the plate itself and the steel bar surface. See Fig. 8.4.

This conclusion is supported by applying a similar three dimensional reconstruction procedure to the other samples. It was not always possible to find a common crossing point, thus indicating that the emission came from different regions. A further reason is that all the paths are verified to start from the sample bar surface and, what is more important, they all came from a surface region close to the tangent line between the steel rod surface and the CR39 plate, as can see in Fig. 8.2. These paths are crossing the thinnest layers of air and, as a consequence, the alpha particles had a relatively low amount of energy. The detected particles may also correspond to the low-energy tail or the high-energy tail of a larger energy distribution. However, the low-energy hypothesis is less realistic, according to the

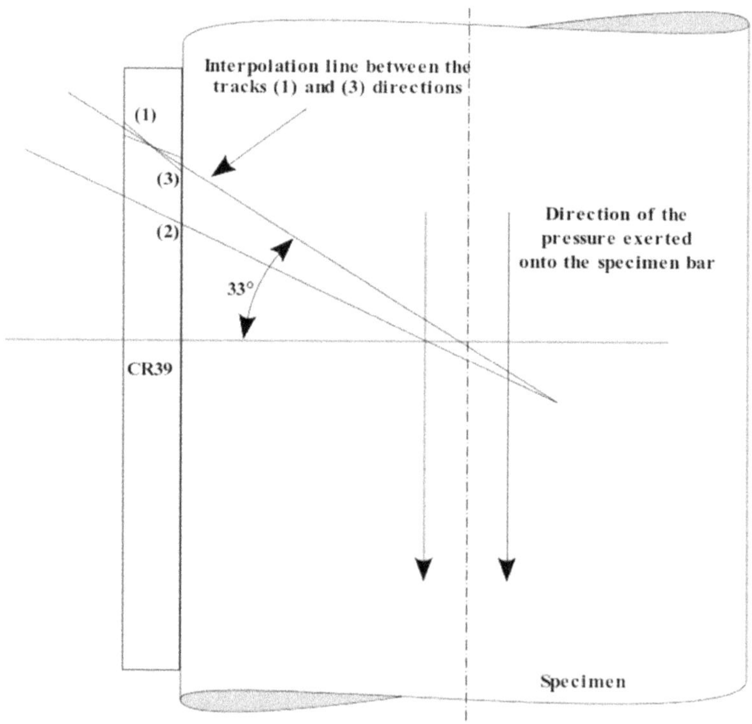

Fig. 8.3. The reconstructed directions for the alpha emission with respect to the horizontal plane and the direction of the pressure exerted onto the bar.

just reported comments. Anyway, for all the tracks, it was possible to evaluate the angle θ between the strain direction and the alpha track direction. In the experimental set-up, the bottom part of the press was kept fixed, while moving the upper part. Thus, the pressure front and the corresponding energy were moving from the top to the bottom as shown in Fig. 8.4. The value of the angle θ so evaluated ranges around 33°, see Fig. 8.3. We want to stress this point in the comparison with the results obtained for the angle of neutron emissions. Moreover this evidence will be further considered in investigating the asymmetry of the neutrons emission angle in Chapter 19.

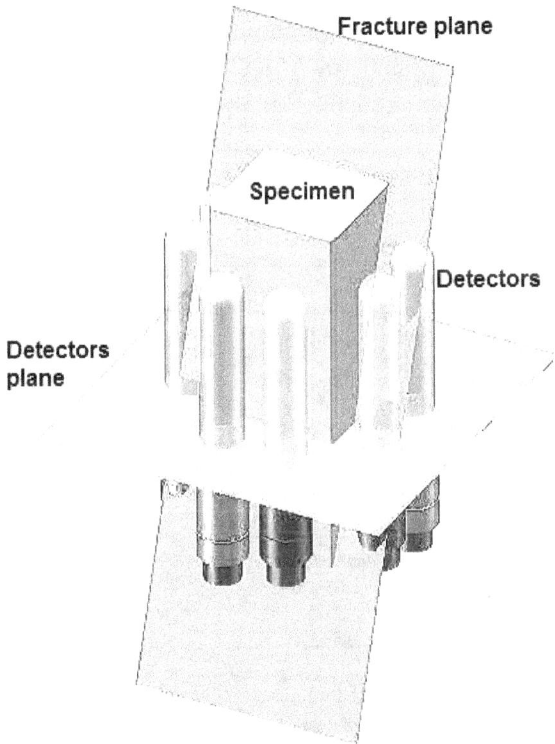

Fig. 8.4. Picture of the specimen with two planes describing the test geometry, the two planes intersect along a line positioned at mid-height and at mid-thickness of the specimen, the two planes are the fracture plane and the detectors plane where the base of the sensitive part of the specimens is located: prospect view.

8.2 Cagliari Experiment: Neutron Emissions

The experiment performed in 2012 at Cagliari was aimed at verifying both the DST neutrons emission during the failure of suitable material undergoing compression to rupture and the direction of the emissions occurring in "bursts" as known by previous experiments, and illustrated in Chapter 6. Pictures of the specimen with two planes describing the test geometry are shown in Figs. 8.5–8.7. The two planes are the fracture plane and the detectors plane where the base of the sensitive part of the specimens is located. The two planes intersect along a line positioned at mid-height and at mid-thickness of the specimen. The angle between the planes is about 63°. The detectors are positioned as in Fig. 8.6. The detector that detected a neutron emission was the C4 as can be seen in Fig. 8.7(a) and for more details in Fig. 8.7(b), it is a thermodynamic neutron

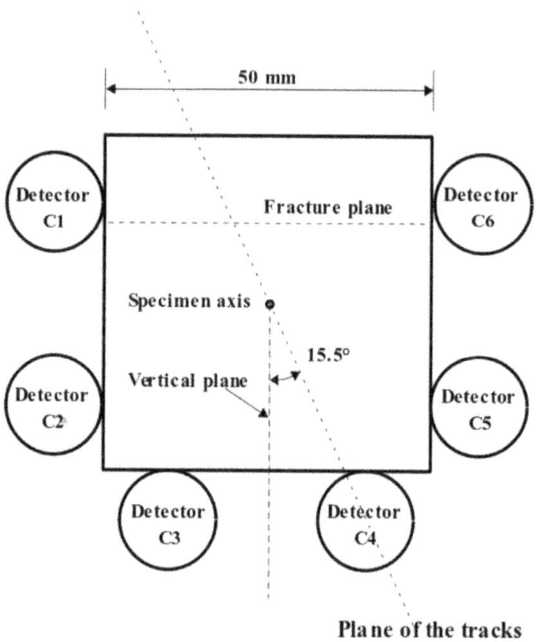

Fig. 8.5. Picture of the specimen with two planes describing the test geometry, the two planes intersect along a line positioned at mid-height and at mid-thickness of the specimen, the two planes are the fracture plane and the detectors plane where the base of the sensitive part of the specimens is located: upper view.

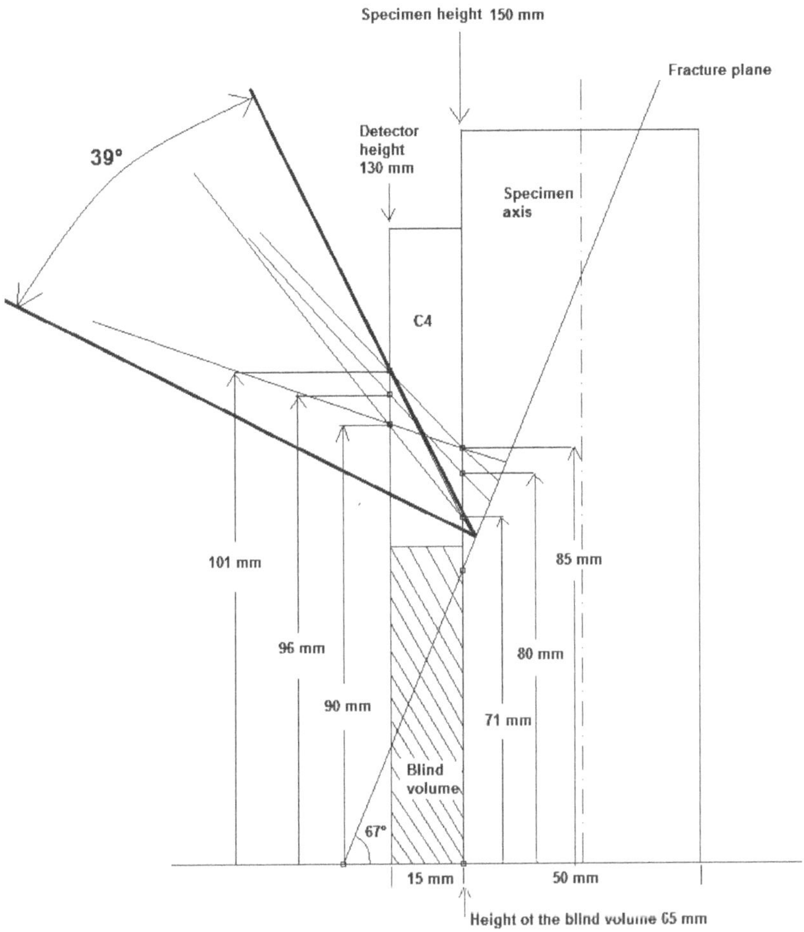

Fig. 8.6. Picture of the specimen with two planes describing the test geometry, the two planes intersect along a line positioned at mid-height and at mid-thickness of the specimen, the two planes are the fracture plane and the detectors plane where the base of the sensitive part of the specimens is located: side view. Reported are the direction of the neutrons emitted and the related angle width.

bubble detector (BD) manufactured by Bubble Technology Industries (BTI), for other details on this kind of detector, see Ref. [4]. It is possible to identify an emission cone of angle $\theta = 39^\circ$ where the neutron bursts occur, see Fig. 8.7. This angle has been reconstructed from the tracks inside the detector as shown in Fig. 8.7 and the geometrical reconstruction scheme shown in Fig. 8.6.

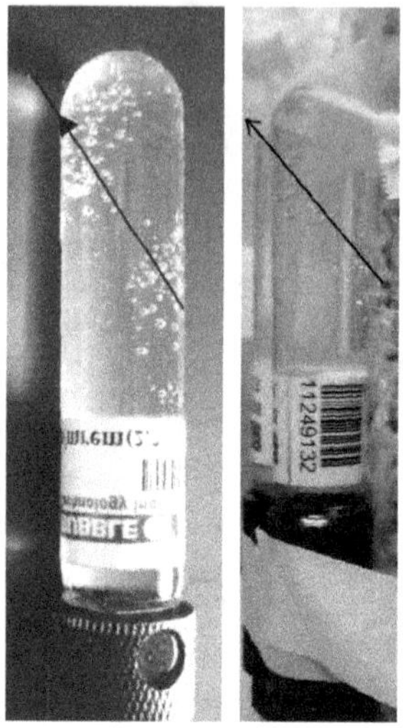

(a) (b)

Fig. 8.7. (a), (b) Neutrons tracks inside the BTI-BD detector C4 showing the direction of emission.

8.3 Remarks on the Results

In the experiments of Ancona for alpha emissions and Cagliari for neutron emissions were identified two main angles for the anisotropic emissions of the particles, and two cases are reported one for each experiment as examples of the behaviour exhibited by all the emissions of each type. In the case of the alpha particle experiments in Ancona the identified angle is about 33°, see Fig. 8.3. In the case of the neutron of Cagliari, the identified angle is about 39°, see Fig. 8.6. We note that within the range of values between these two angles arises the value of the principal angle of 35.3°, corresponding to the direction along which there is violation of Lorentz local invariance as occurred in other experiments which, as already said we will see in the next Chapters 13–15. These results about the nuclear particles emissions are only a suggestion on the fundamental role that the angle of emission has in these processes that violate the Lorentz invariance. We therefore think that the violation of Lorentz invariance at the base of the DST emission processes, and in the more general context of the DST reactions, is intimately connected to an angle of asymmetry of emissions both of alpha and neutrons. We therefore consider that there is an angle of breakdown for the Lorentz invariance which must be searched for in all experiments where the Lorentz invariance is violated. We intend to present the results of this survey in a subsequent Chapters 19–21 which have the purpose of connecting the different fundamental interactions with the respective angles of violation of Lorentz invariance. We suppose, in fact, that it is not sufficient to take into account the anisotropy in the geometric representation of the interactions but must also take into account the asymmetry. However in the context of DST reactions, it has been performed an experiment of 'stereoscopic' measure of DST emissions, in particular neutrons at least on a plane surface, as we have seen in Section 6.2.

A stereoscopic measure of DST neutron and of their energy gave certainly a clear and unequivocal indication of the fundamental role that the angle of asymmetry plays in nuclear reactions in which Lorentz invariance is violated as we will discuss in Chapter 19.

References

[1] G. Albertini *et al. Mater. Process Energy,* ed. by A. Mendez-Vilas, Ed. Formatex 2013.
[2] G. Albertini *et al. J. Adv. Phys.*, 2(1), 20–24, 2013.
[3] F. Cardone and S. Duro. *Modern Phys. Lett. B*, 28(19), 1450156–1, 2014. DOI: 10.1142/s0217984914501565.
[4] A. Manuello *et al. Energy Emission from Brittle Rocks under Mechanical Load,* Meccanica, Springer Nature Ed. p. 1177–1188 (2015) DOI: 10.1007/s11012-014-9987-9.

Chapter 9

Proton Antiproton Annihilation

We try now to explore the 'shape' of energy as it appears in the phenomenon of the annihilation. In Sections 1.4 and 2.4, we show that the hadronic metric in four dimensions has been obtained by the analysis of the correlation among the pions produced by proton antiproton annihilation. It is possible to perform a pseudo – four dimensional analysis of the 'form' assumed by the energy coming from the annihilation and producing pions as last stage of the so called process of hadronization. In literature such a 'form' of the annihilation energy is named fireball. The values in centimetres, seconds and in GeV are reported in Table 9.1 for colliding energies above the hadronic threshold $E_{0\mathrm{had}}$ and in Table 9.2 for colliding energies below $E_{0\mathrm{had}}$. The form of the fireball results in a sort of ellipsoid and therefore we call semi – axes the physical dimensions along each coordinate axis. We recall that the axis z is that one along which the proton and antiproton collide.

Note the fact that the a_i values reported in Table 9.2 can be assumed valid as well, even for electron-positron annihilation for center of mass energies below $E_{0\mathrm{had}}$, with the condition that axis z coincides with the collision direction of the electron and positron. We use the values reported in Table 9.2 to perform the following representation assuming a new point of view of the homogeneity concept. In general after Minkowski convention it is usual to consider the coordinates expressed in spatial units of measure. This is the classical concept of homogeneity, and therefore people look for invariant constant, in the Einsteinian mechanics sense, giving this conversion in spatial units.

Table 9.1. Hadronic physical parameter a_i of the fireball for proton and antiproton center of mass energies $E \geq E_{0\text{had}}$, for E the exponent k ranges from 1 to 6 accordingly if the ratio $(E/E_{0\text{had}})$ is nearly 1 or very greater than 1.

Reference frame axis	Semiaxes a_i $i = 1, 2, 3, 4, 5$	Values
x	a_1	$0.06\ 10^{-13}$ cm
y	a_2	$0.08\ 10^{-13}$ cm
z	a_3	$0.20\ 10^{-13}\ (E/E_{0\text{had}})$ cm
t	a_4	$0.67\ 10^{-24}\ (E/E_{0\text{had}})$ sec
E	a_5	$0.93\ (E/E_{0\text{had}})^{\text{k}}$ GeV

Table 9.2. Hadronic physical parameter a_i of the fireball for proton and antiproton center of mass energies $E \leq E_{0\text{had}}$, where in coordinate E the (m_p) stands for proton mass corresponding to the value assumed by a_5.

Reference frame axis	Semiaxes a_i $i = 1, 2, 3, 4, 5$	Values
x	a_1	$0.06\ 10^{-13}$ cm
y	a_2	$0.08\ 10^{-13}$ cm
z	a_3	$0.20\ 10^{-13}$ cm
t	a_4	$0.67\ 10^{-24}$ sec
E	a_5	0.93 GeV (m_p)

Table 9.3. Non rotation ellipsoid semiaxes in GeV (see Fig. 9.1).

Coordinate axis	Ellipsoid semiaxes	GeV
x	a_1	3.28
y	a_2	2.46
z	a_3	0.99

We did the same in previous chapters and in particular introducing the Pessa constant in Par.4.3. Now we assume the point of view that homogeneity only requests the same physical unit of measure and not a specific or privileged one. Therefore we apply the homogeneity concept assigning to the semiaxes a_i the dimension of energy. To this aim we use the values from Table 9.2 and, since we

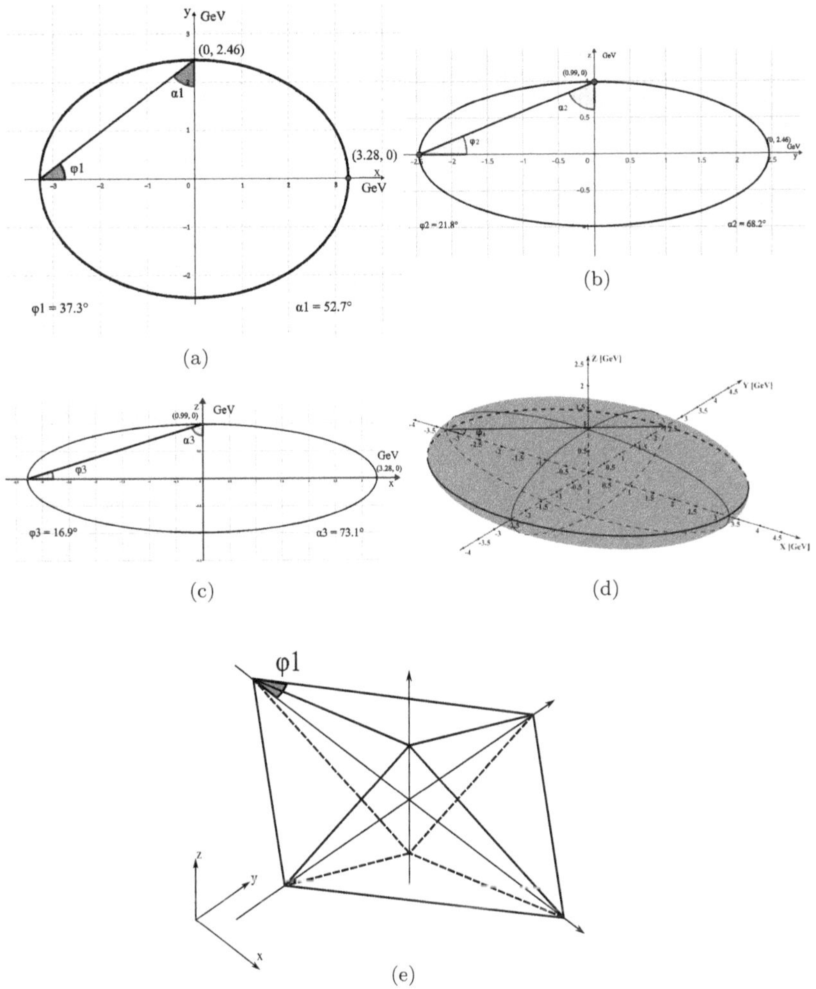

Fig. 9.1. Orthographic projections (a), (b), (c), prospect view (d). (e) Cauchy tetragon inscribed in the ellipsoid shown in (d).

are in Minkowskian conditions, flat space time, we use the following conversion relations assuming as energy unit the electron Volt [eV]:

$$a_k[eV] = \hbar\frac{c}{a_k}[cm] \quad k = 1, 2, 3; \quad a_4[eV] = \frac{\hbar}{a_4}[sec]; \quad a_5[eV]. \quad (9.1)$$

We remind that h and c are invariant in Einsteinian mechanics and Minkowskian space time. The geometrical results are sketched in Figs. 9.1 and 9.2, for the sake of clarity in two and three dimensional

Table 9.4. Rotation ellipsoid semiaxes in GeV (see Fig. 9.2).

Coordinate axis	Ellipsoid semiaxes	GeV
x	a_1	3.28
z	a_3	0.99
t	a_4	0.98

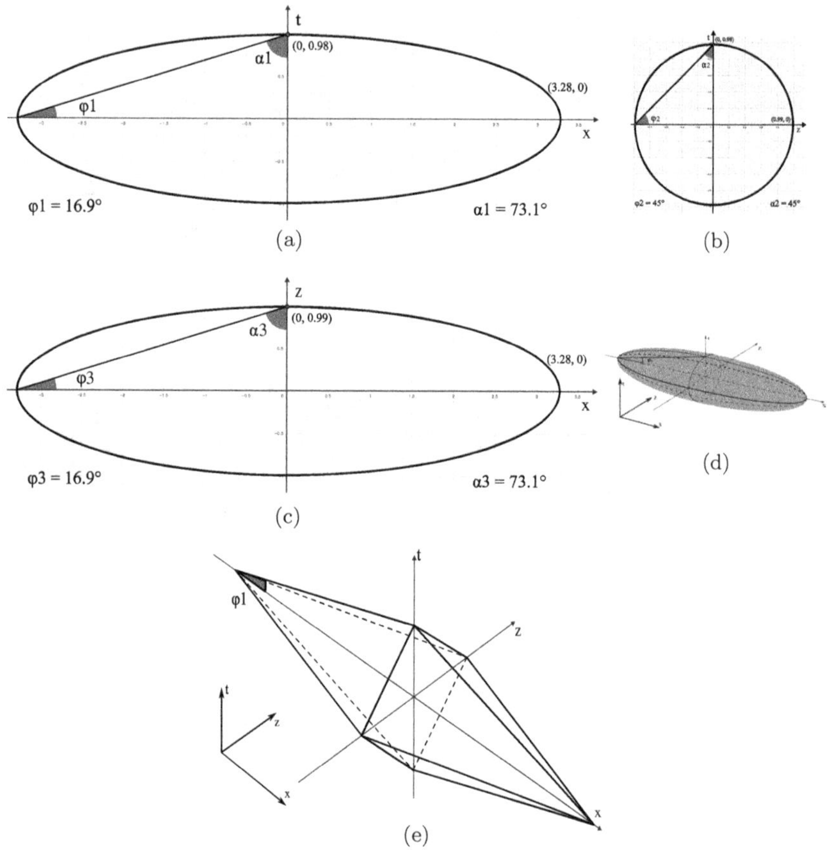

Fig. 9.2. (a) Ellipsoid section in the plane x, t. (b) plane z, t. (c) section in the plane x, z. (d). (e) Cauchy tetragon inscribed in the ellipsoid shown in (d).

representations. In Fig. 9.1 we can see that the resulting shape of the 'volume' of the energy is a non – rotation ellipsoid or rather having all the semi – axes different and it holds for the following three coordinate systems: (xyz); (xyt); (xyE). Instead in Fig. 9.2 the resulting shape is a rotation ellipsoid having two semi – axes equal and it holds for the following six coordinate systems: (xzt); (xzE); (xtE); (yzt); (yzE); (ytE).

We suggest considering Figs. 9.1 and 9.2 as two and three dimensional 'projections' or 'sections' of a physical reality that is actually five dimensional.

The further considerations is about the fact that in Fig. 9.2 the two equal semiaxes occur ever in the couples of coordinate axes: (zt); (zE); (tE). We borrowed from the theory of elasticity the concept of Cauchy tetragon, used to make up the stress tensor. We sketched within each ellipsoid a tetragon made by connecting all the semi-axes ends and moreover reporting the angles, in Euclidean geometry, formed with the coordinate axes, see Figs. 9.1(e), 9.2(e). In Table 9.3 we report the values of the semiaxes in GeV which refer to the non rotation ellipsoid shown in Fig. 9.1; in Table 9.4 there are the values which refer to the rotation ellipsoid shown in Fig. 9.2.

Chapter 10

Symmetry Recovered by Energy

We present now the representation of the energy 'volume' in the reference frame (z t E). We remind that axis z is along the direction of matter antimatter collision. As noted in the previous chapter we will see that the representation shall be symmetrical. In Table 10.1 we report the values of the semiaxes in GeV and in Fig. 10.1 is sketched the result which is a spheroid since the semiaxes are nearly equal and therefore the Cauchy tetragon is nothing else than a cuboid.

It is intriguing enough that a suitable choice of coordinate axes together with a suitable interpretation of the homogeneity request with the choice of energy as the same unit of measure give rise to the possibility of a geometrical representation of the energy having the property to be symmetrical or near-symmetrical as a spheroid with low eccentricity. Now let us make some more considerations about this possibility to recover a lack of symmetry by using the energy symmetrization. First of all we note that representing the 'volume' of energy we move from a geometrical entity, the non rotation ellipsoid, having high eccentricity, see Fig. 9.1, through a rotation ellipsoid with lower eccentricity, see Fig. 9.2, then to a spheroid with lowest eccentricity, see Fig. 10.1. We have to take into account that we followed this way starting from the hadronic metric having the property of anisotropy independently by the energy and therefore showing a violation of the local Lorentz invariance (LLI). The anisotropy of hadronic metric brings to a lack of symmetry, consequence as well of the lack of LLI. Now it is possible to state that a violation of LLI can be recovered introducing a new physical extended coordinate and

Table 10.1. These data refer to Fig. 10.1.

Coordinate axis	Semiaxes	GeV
z	a_3	0.99
t	a_4	0.98
E	a_5	0.93

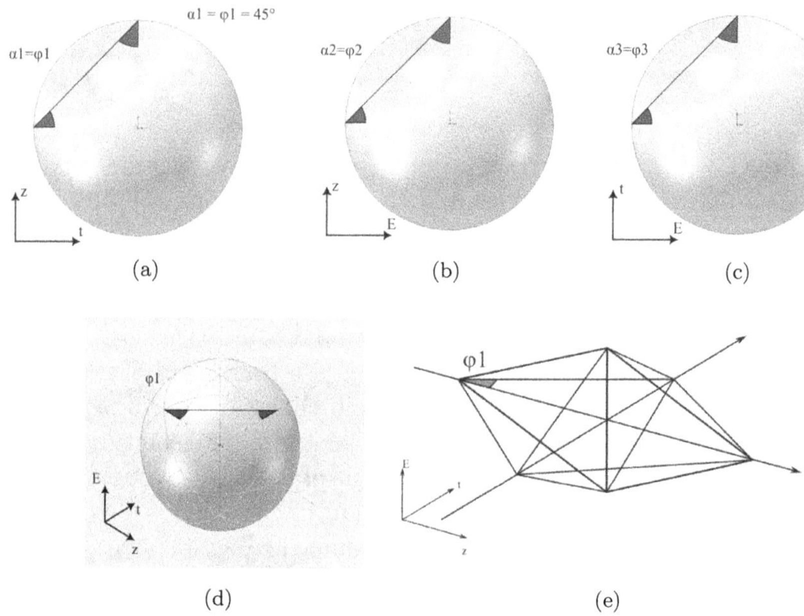

Fig. 10.1. In (a), (b), (c) are reported the orthogonal section of the spheroid according canonical planes zt, zE and tE. In (d) the three dimensional sketch of the spheroid. In (e) the Cauchy tetragon approximating a cuboid.

perform a representation of physical phenomena by a metric space endowed with one more coordinate. As it was already done when the Galilean symmetry was lost in electromagnetic phenomena: it was useful, or better necessary, to introduce one more coordinate, the time, and to change the Euclidean three dimensional geometry, giving the Galilean symmetry of the translations, with the Minkowskian four dimensional geometry, giving the LLI or rather the Lorentzian symmetry of the roto-translations. In our cases under consideration the violation of LLI represented by anisotropic metrics, if even asymmetric, is recovered by introducing a new coordinate, the energy, with a five dimensional geometry. In order to recover the symmetry we are following the master way to get the symmetrization of energy distribution in space and moreover with respect to the angular distribution. As we will see later this way can be followed better by experiments than mathematics, at least up to now. About the issue of the angular distribution of energy in the phenomena, we did the choice to borrow from the theory of elasticity the concept of Cauchy stress tetragon in order to evaluate the notable angles connected with hadronic metric which we will use in the next. But more in general we will see from phenomenological evidences that the angle will assume a more and more relevant role in describing both the lack of symmetry and the lack or violation of LLI as far as being connected with LLI itself.

Chapter 11

Critical Volume

11.1 Cavitation Bubble Dimension

What we call 'cavitation' has to be considered synonymous of 'spherical symmetrical collapse' of a gas bubble, and in this sense, we are going to use this term 'cavitation' from here on. Moreover, in order to avoid any misleading with the known nuclear reactions, such as fusion and fission of nuclei, we call the deformed space time reaction among nuclei by new terms: nucleosynthesis, when parent nuclei form a composite nucleus, and nucleolysis, when a parent nucleus splits into other nuclei.

Regarding the way in which cavitation can produce the energies needed to induce DST reactions, let us take into consideration some of the physics underlying the cavitation process. It consists of the implosive collapse, under suitable pressure conditions, of a gas bubble within a liquid or a gas bubble inside the porosity of a solid. We start to consider water in order to make predictions on suitable bubble size in the framework of DST theory. The speed of sound is about $v \approx 10^3$ m/s which, on account of the used frequency $\sim 10^4$ Hz, corresponds to a wavelength $\lambda \sim 10^{-1}$ m. In order to get the gas bubble to implode, the plane pressure wave of the ultrasounds must be converted into a symmetric spherical shock wave on the bubble surface. Therefore, the condition to be satisfied for an implosive, spherical, symmetrical collapse is that the wavelength must be much greater than the bubble size. Taking as example a spherical bubble with radius R; this means $\lambda \gg R$. In general, at room temperature and atmospheric pressure, the size of a bubble subjected to cavitation is expected to be classically of the order of magnitude $R = 10^{-6}$ m [1,2].

We now remark that the only atoms influenced by the shock wave producing the cavitation are the ones lying on the surface of the bubble itself.

These atoms are trapped in a double layer volume at the border between the bubble material and the external material. They are stuck by the surface tension of the bubble which is generated by the combined electrostatic repulsion of the bubble material and the surrounding external material.

Namely, all other atoms inside the bubble volume escape to the outside during the collapse, due to the fact that the inner pressure, corresponding to the saturated vapour pressure, is far lower than the external pressure. Moreover, the inter-molecular gap in the external surface is greater than the atomic dimensions, giving space for the inner atoms to escape. Actually, the collapsing bubble behaves more as an 'accelerator' than a piston, in this sense the collapse works as a micro reactor. The trapping of atoms at the bubble surface is expected to be more effective for metals. Indeed, it is known for a solution that usually metals and metal cations do not enter the cavitation bubble but stay at the interface between bubble and bulk solution. In the case of aqueous solutions at room temperature, 300 K, the saturated vapour pressure of water is 0.02 bar, whereas we estimated that the pressure induced by the implosive shock wave, for a transmitted pressure power of 100 W, is of the order of 10^9 bar. This circumstance entails that there is no limit to the spatial size attainable by the collapse. It is therefore reasonable to hypothesise that the lower limit of this size can be identified with the nuclear size. As a matter of fact, there is in literature no experimental information on the value of the minimum size attained by a collapsing cavitation bubble. Therefore, we can suppose that at the end of collapse, the bubble dimensions become near to the nuclear dimensions, commonly about 10^{-15} m.

If the wavelength λ of the plane pressure wave satisfies the condition $\lambda \gg R$, where R is the bubble radius, the plane pressure wave becomes a spherical shock wave symmetrically acting on the bubble according to the Pascal principle, see Fig. 11.1.

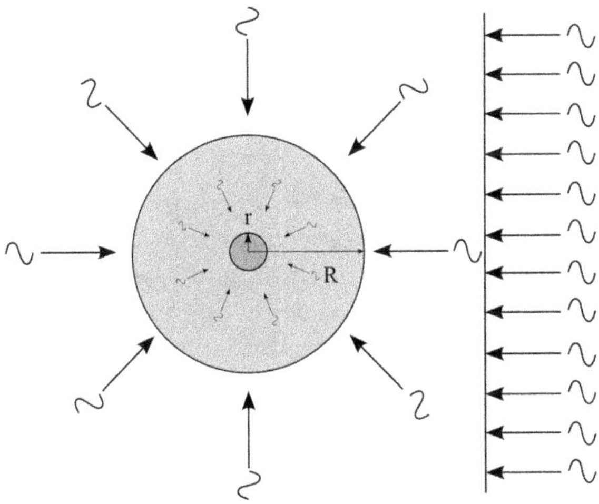

Fig. 11.1. Conversion of a plane pressure wave into a spherical pressure wave on the bubble surface during cavitation process.

Let P be the power of the plane pressure wave, and R, r the radius of the bubble before and after collapse, respectively. Then, the power densities on the bubble before and after collapse are $D_P = \frac{P}{4\pi R^2}$ and $D'_P = \frac{P}{4\pi r^2}$. If the initial energy flux, the energy for unit time and unit surface, is conserved due to the continuity equation and to energy conservation, we have

$$P = SD_P = (4\pi r^2)D'_P \Rightarrow D'_P = D_P\left(\frac{S}{4\pi r^2}\right) = fD_P, \qquad (11.1)$$

where $f = \frac{S}{4\pi r^2}$ is the amplification factor due to the initial surface S and the collapse end surface.

As a consequence, the power density on the bubble surface, generated by the power of the plane pressure wave, produces after implosion a power density increased by the factor f on the reduced bubble surface. Such a factor, caused by the implosion, ranges from $f_A \sim 10^8$ when the bubble collapses to atomic size, $r_A = 10^{-8}$ cm, up to $f \sim 10^{18}$ collapse to nuclear radius, $r_N = 10^{-13}$ cm. In the experiments have been used powers ranging from 100 W to 630 W, or rather from 6×10^{20} to 4×10^{21} eV/s, so the final power density is of the order of $10^{45} - 10^{46}$ eV/(s $\times$ cm^2). By the way, classically, for a low-density plasma, this power density corresponds to an equivalent temperature of at least 10^{20} K. But we do not follow this classical point of view since this temperature, even at microscopical level, cannot be considered realistic. Actually, as we already noted, the collapsing bubble acts more as a micro 'reactor' than a piston. When the bubble collapses and the bubble surface shrinks to the nuclear dimensions, trapped atoms do move together with the surface and come closer and closer. Therefore, the collapsing surface of the bubble behaves as an 'inertial accelerator' of neutral atoms and the squeezing of the surface to the nuclear dimension produces on the bubble surface the energy density required to activate the DST reactions. At the final stage of this process, atomic electrons are stripped away.

It is possible to set as well a classical model of the reaction catalyzed by bubble collapse, but here we want to point out some limits of any classical model.

First of all, it must be noted that not all of the power from the sonotrode is concentrated in one bubble of nuclear dimension. Also, much of the power put in is dissipated in other competitive processes, say endothermic sonochemical reactions. This could be taken into account by an approximated energy efficiency factor, useful for non-sonochemical reactions, the latter are to be identified with those reactions that involve parent nuclei to generate composite nuclei with a greater mass. Moreover, a crucial parameter is just represented by the final size reached by the imploded bubble, which critically determines the energy available for any nuclear process. Another intrinsic limit of the classical model is due to the very phenomenology of nuclear reaction of heavy nuclei, when we consider the Minkowskian case of nuclear fusion instead of the DST case of the nucleosynthesis. At energies below the barrier, the fusion probability is low and rises exponentially with energy. At energies just above the Coulombian barrier, the probability for parent nuclei to fuse and form a composite nucleus does increase further but not indefinitely. As a matter of fact, at energies of about 10 MeV per nucleon, the cross section for forming a composite system is low and diminishes at a rate of $1/E_{\mathrm{cm}}$, where E_{cm} is the centre-of-mass energy of the relative motion of the two nuclei. This decrease occurs because the composite system formed at such high relative energies is a highly excited nucleus and has large angular momentum. This causes the composite system to fission nearly instantaneously, with a time frame much shorter than 10^{-22} s.

A possible answer to both these questions may come from the hadronic metric whose main features we recall in the following:

$$g_{\text{had}}(E) = \text{diag}(b_{0,\text{had}}^2(E), -b_{1,\text{had}}^2(E), -b_{2,\text{had}}^2(E), -b_{3,\text{had}}^2(E)),$$

$$\left.\begin{aligned} b_{1,\text{had}}^2(E) &= \left(\tfrac{\sqrt{2}}{5}\right)^2 \\ b_{2,\text{had}}^2(E) &= \left(\tfrac{2}{5}\right)^2 \end{aligned}\right\} \forall E > 0, \tag{11.2}$$

$$b_{0,\text{had}}^2(E) = b_{3,\text{had}}^2(E) = \begin{cases} 1, & 0 < E \leq E_{0,\text{had}} \\ (E/E_{0,\text{had}})^2, & E_{0,\text{had}} < E \end{cases}$$

$$= 1 + H(E - E_{0,\text{had}})\left[\left(\frac{E}{E_{0,\text{had}}}\right)^2 - 1\right], \quad E > 0, \tag{11.3}$$

where $H(x)$ is the Heaviside function and

$$E_{0,\text{had}} = (367.5 \pm 0.4)\,\text{GeV} \tag{11.4}$$

is the hadronic threshold energy. $E_{0,\text{had}}$ is the energy value at which the metric parameters are constant, and the metric becomes flat Minkowskian. Note again that the hadronic metric is over-Minkowskian, namely, it approaches the Minkowskian flat space time limit from above ($E_{0,\text{had}} < E$). Let us recall that, as described, the power density attained in the mechanism of the bubble collapse to the nuclear radius is $D_{Pn} \sim 10^{45} - 10^{46}$ eV/(s cm^2). In order to find the energy achieved in this case, we have to estimate the collapse time of the bubble. The estimated [1] initial bubble radius is $R \sim 10^{-6}$ m, whereas the nuclear radius is $r \sim 10^{-15}$ m, giving the surfaces initial and final for the continuity relation Eq. (11.1). For an employed frequency $\nu = 20$ kHz, the period is $T = 1/\nu = 5 \times 10^{-5}$ s. The collapse time is obviously comprised between the upper limit T and the time $t_s = R/v_s$ taken by the pressure wave to travel the bubble radius at the sound speed: $t_s < t_c < T$. In deionized and bidistilled water, the sound speed is $\nu_s \sim 1.4 \times 10^3$ m/s. This corresponds to the lower time limit for collapse $t_s = 7 \times 10^{-10}$ s.

The collapse energy E_c corresponding to the power density D_{Pn} attained for collapse to the nuclear radius is therefore

$$E_{\text{cmin}} = SD_{Pn}t_s < E_c < E_{\text{cmax}} = SD_{Pn}T. \tag{11.5}$$

We get $E_{\text{cmin}} \sim 10^8$ MeV; $E_{\text{cmax}} \sim 10^{13}$ MeV.

Let us note some facts. The energy E_{cmin} is much higher than the maximum estimated Coulomb barrier for the intermediate mass nuclei. Moreover, the range of values of E_c corresponds to the energies of the direct nuclear reactions induced by photons and to the production of pairs e^+e^- due to photon conversion in the nuclear electric field.

The important point is that E_{cmin} is also much higher, by three orders of magnitude, than the hadronic threshold of the hadronic interaction metric, $E_{0,\text{had}} \sim 4\times 10^5$ MeV, see Eq. (11.4). This implies that the hadronic interaction between nuclei occurs in the non-flat part of the hadronic interaction metric. Therefore, hadronic spacetime geometry is deformed in the final collapse region. Obviously, for $E \gg E_{0,\text{had}}$, local Lorentz Invariance (LLI) is violated. As already noted, in the usual flat Minkowskian metric, the interacting nuclei, although overcoming the Coulomb barrier, produce, after fusion, a nucleus in an highly excited state and therefore with high probability of spontaneous fission. On the contrary, in the deformed space time produced by the over threshold hadronic condition, the excess energy after fusion goes into deforming the spacetime region, thus leaving the son nucleus, produced by the nucleosynthesis of the two parent nuclei in a low-excited or even unexcited state. As we showed, the stability of the nuclei produced by cavitation is due to the deformation of both space and time in the collapse region ensuing from the non-Minkowskian behaviour of the hadronic interaction in the range $E_c > E_{0,\text{had}} = 3.675 \times 10^{11}$ eV. For such energy values, local Lorentz invariance is violated for hadronic nuclear interaction. This allows us to determine an energy threshold for the occurrence of stable DST reactions induced by cavitation.

The time deformation law in a hadronic field reads

$$\frac{dt_{\text{had}}}{dt_{\text{e.m.}}} = \frac{E_{0,\text{had}}}{E}. \tag{11.6}$$

A way to read Eq. (11.6) is as follows. It can be regarded as an action–reaction relation, or better as the equality between two energy speeds.

A speed $W_{\text{e.m.}}$ of supplying energy to the atoms by the electromagnetic interaction, the action, and a speed W_{had} of the response by the hadronic interaction of nuclei, which is the reaction

$$W_{\text{had}} = E_{0,\text{had}}/dt_{\text{had}} = E/dt_{\text{e.m.}} = W_{\text{e.m.}}. \tag{11.7}$$

The interpretation of Eq. (11.6) and its use in Eq. (11.7) is fully understood by recalling the role of energy as a fifth dimension.

In order to get the threshold of LLI breakdown for hadronic nuclear interaction, the time taken by a generic cavitating bubble to collapse for a given electric energy E must supply an energy speed $W_{\text{e.m.}}$ so as to equate the nuclear one. Let dt_{had} be the nuclear reaction time given by

$$dt_{\text{had}} = \gamma_{\text{had}}\Delta t, \tag{11.8}$$

where $\Delta t = h/m_\pi$ is the Yukawa time, the so-called 'nuclear year'.

To get an estimate of dt_{had} at the energy threshold $E_{0,\text{had}}$, one can use the relation $\gamma_{\text{had}} = E_{0,\text{had}}/m_\pi$, on account of the well-known fact that $\gamma = E/m$ is the relativistic factor of time dilation in Minkowskian conditions for $E \leq E_{0,\text{had}}$. Replacing such an expression of γ_{had} in Eq. (11.8) yields

$$dt_{\text{had}} = (h/m_\pi^2)E_{0,\text{had}}, \tag{11.9}$$

where $h = 4.136 \times 10^{-15}$ eV s, $m_\pi = (m_{\pi\pm} + m_{\pi 0})/2 = 1.373 \times 10^8$ eV,

$$E = dt_{\text{e.m.}} E_{0,\text{had}}(m_\pi^2/E_{0,\text{had}}h) = dt_{\text{e.m.}} W_{\text{had}}. \tag{11.10}$$

Since $E = dt_{\text{e.m.}}(m_\pi^2/h)$, W_{had} reads

$$W_{\text{had}} = m_\pi^2/h = 4.8 \times 10^{30} \text{ eV s}^{-1} = 7,6 \cdot 10^{11} \text{ W}. \tag{11.11}$$

Let us assume for $dt_{\text{e.m.}}$ the time taken by a microbubble of radius R to collapse to the nuclear size with $r \sim 10^{-13}$ cm due to the electric repulsion of the water surrounding atoms subjected to the ultrasonic pressure wave. The collapse can occur at the velocity of sound in distilled water, $v = v_s = 1.4 \times 10^3$ m/s, or at the velocity of the shock wave, $v = v_u = 4v_s$. Since the ultrasound wavelength is much greater than the microbubble diameter, it is $dt_{\text{e.m.}} = R/v$ in either case.

Table 11.1. Bubble radius and related energy thresholds for sound speed collapse.

$v = v_s$			
R_{bubble}(m)	10^{-6}	2×10^{-6}	4×10^{-6}
E_{thres} (J)	5×10^2	10^3	2×10^3

Table 11.2. Bubble radius and related energy thresholds for shock speed collapse.

$v = v_u = 4v_s$			
R_{bubble}(m)	10^{-6}	4×10^{-6}	8×10^{-6}
E_{thres} (J)	10^2	5×10^2	2×10^3

Therefore, we have, for the reaction threshold energy E_{thres},

$$E_{\text{thres}} = (R/v) \cdot (m_\pi^2/h). \tag{11.12}$$

Tables 11.1 and 11.2 summarise the values of E_{thres} with varying radius of collapsing microbubbles for the two possible collapse speeds v_s and v_u, deduced from Eq. (11.12).

In order to produce stable DST reactions, and therefore a stable DST emission of particles, it is necessary to supply constantly by pressure to the system of distilled water and solute an energy $E \geq E_{\text{thres}}$. Such a condition permits to triggering of DST reactions in the presence of broken local Lorentz invariance. By using a cavitator absorbing 2000 W and able to provide a stable supply from 100 J up to a maximum of 2000 J, it is possible to investigate the collapse of bubbles with sizes ranging from 1 μm to 8 μm, by taking either v_s or $v_u = 4v_s$ as the collapsing speed. However, the experiments carried out with energies of the order of 100 J have provided evidence for the occurrence of DST reactions both in liquids and in solids, see Chapters 6 and 7. For the liquids, water solution, this is in favour of the velocity of the shock wave as collapsing velocity, and therefore of the model in which the ultrasound pressure plane wave generates a symmetrical, spherical shock wave around the bubble. Ever about the liquids nothing can be said about the total mass of water and compound to be cavitated, or about the amplitude of the ultrasonic wave. Both of them are phenomenological parameters and have been determined empirically. Needless to say, a higher supplied energy corresponds to a greater amplitude and a lower energy is available for cavitation and bubble collapse if a greater mass is subjected to ultrasounds.

As we have seen, the existence of the threshold for DST reactions E_{thres} is a direct consequence of the existence of the energy threshold $E_{0,\text{had}}$ for the hadronic interaction. This circumstance allows one to discriminate between signals coming from the nuclear reactions produced by cavitation in general and the DST reactions. In fact, if the nuclear particles emitted by the nuclei after interaction are alphas and neutrons, the son nuclei are left in an excited rotational state. Therefore, due to angular momentum conservation, they decay to a lower state by emitting γ radiation. Then, from a classical viewpoint, such a disexcitation process is accompanied by both nuclear particles and photon radiation. However, if the interaction among nuclei occurs in non-Minkowskian conditions, for metric conditions ruled by $E > E_{0,\text{had}}$, the excess energy is partly absorbed by the hadronic space time deformation so that there is no emission of γ radiation.

The two facts of the energy metric overcoming the threshold, $E > E_{0,\text{had}}$, and of the alpha and neutron emission in the absence of γ radiation do provide the complete signature of DST reactions produced by the cavitating collapse of gas bubbles inside condensed matter in DST conditions.

In order to test this, it was enough to check that in condensed materials, liquid or solid, subjected to cavitation, no emission of neutrons occurs without emission of γ. Or rather, even at the energy reaction threshold $E_{\text{thres}} = E$, we have the usual situation: neutrons together with gammas. This means that the DST reaction conditions were not achieved anyway.

By measuring the neutrons produced by the DST reactions generated by cavitation at or over threshold, it will be possible to determine the radiative calorimetry of the produced process, see Section 6.2.

11.2 The Ridolfi Cavities

Now, we test the prediction of Tables 11.1 and 11.2 inside condensed matter. Up to now, we used for liquid matter, water and water solution, the estimate of the order of magnitude, usually reported in literature, for minimal size of collapsing bubbles. But the size of the micro reactor, bubble in liquid, cavity in solid, ranges in the typical interval 4–8 micron. Unfortunately, it is very difficult 'to see' gas bubble size inside a liquid, even by echo Doppler devices or by laser diffractometry. On the contrary, in a solid, such as iron or steel, it is possible to look and evaluate its porosity, by 'seeing' the cavities and their size inside the bulk of the material. The systematic study of the iron, steel and ferrite bars giving rise to neutron emission, as in Chapter 6, gives us the tool to verify this prediction.

The ferrite represents an iron and carbon alloy with superior purity and with a decidedly inferior carbon content with respect to the steel. Sample preparation consisted of radial extraction of two parallelepipeds from the cylindrical ferrite bar using a wire Electric Discharge Machining (EDM). The long dimensions of the parallelepipeds correspond to the bar diameter, 20 mm, and the square, curved bases have an area of $\sim$4 mm^2. The base of the first parallelepiped contains one of the dark, damaged areas, Figs. 11.2(a) and 11.2(b), whereas the second sample was extracted from an undamaged zone of the bar surface [3, 4].

A first series of analyses of the square bases of the two parallelepipeds representing the external curved surface of the bar were performed using optical microscopes and an Environmental Scanning Electron Microscope (ESEM). Most of the ESEM analyses were performed by a FEI Quanta 200 which is equipped with a field emission gun, secondary electron (SE) and back scattered electron (BSE) detectors, and an Energy Dispersive System (EDS) for X-ray atomic analysis. Microscope and ESEM observations allowed us to obtain microstructural data of the external damage and the interior of the ferrite bar, see Figs. 11.2 and 11.3.

Back Scattered Electron (BSE) imaging of the damage mark shows a cratered surface of lower brightness, low average atomic weight, with respect to the intact ferrite surface, see Figs. 11.2(b) and 11.2(c). Bright remnants of the original ferrite surface can also be found between the cratered zones, which are characterized by irregular and cracked interiors composed of a dark material with a lower mass, see Figs. 11.2(c)–11.2(e). The cross-sectional view of the damage mark indicates that the thickness of the low-brightness material is minimal, possibly $\leq$1 micron, see Fig. 11.2(f). The deeper

Fig. 11.2. Microphotograph (a) and BSE images (b)–(f) of the parallelepiped with the damage mark. (a) and (b) Panoramic views of the damaged surface; an oxidized patch next to the damage mark is also shown. (c) and (d) Close-up views of the damage mark. (e) An enlargement of (d) showing a deep crater surrounded by a dark and cracked surface. (f) A cross-sectional view of the damage mark; the occurrence of small cavities is also shown.

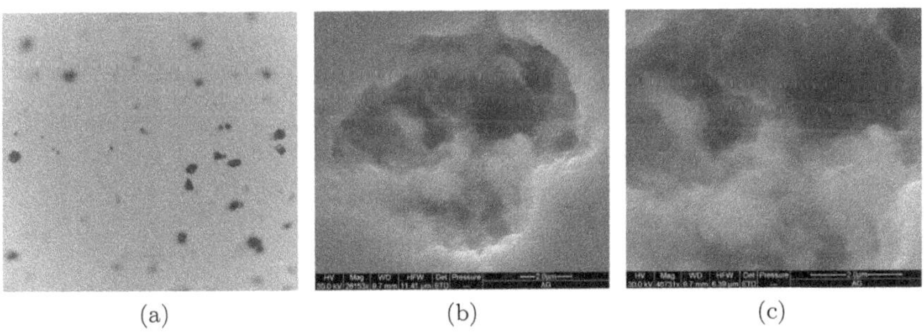

Fig. 11.3. Microscope polarized light (a) and SE images (b)–(c) showing a microcavity-rich zone inside the bar (a) and the microstructure of one of these cavities (b). (c) A close-up view of the cavity interior in (b).

craters have depths up to 20 μm and, together with the bright remnants of intact ferrite, suggest that highly energetic microexplosions occurred, see Figs. 11.2(e) and 11.2(f).

The internal structure of the bar appears mostly uniform with occasional defects in the form of roundish microcavities, Fig. 11.2(f). However, a zone of $\sim$0.5 mm^2 with a large number of microcavities, $\sim 1.3 \times 10^3$ per mm^2, was found at a distance of $\sim$0.3 mm from the external damage mark, see Fig. 11.3(a). The largest of these cavities has a diameter of $\sim$10 μm. At higher magnifications, the cavities appear highly deformed with irregular contours and are partially filled with chaotic, amorphous material, see Figs. 11.3(b) and 11.3(c). In addition, this material appears stable under the heating effect of the focused electron beam.

It is worth noting that neither the external damage zones nor the areas characterized by a high density of microcavities filled with amorphous material were found on either the external surface or the interiors of the reference parallelepiped, which did occasionally show smaller-sized, <5 μm, empty cavities. The fact that below 5 micron size of cavities there are no inner spurious materials is at the very base of the test for the predictions in the previous section, see Tables 11.1 and 11.2.

The Energy Dispersive System (EDS) analyses of the ferrite matrix and the anomalous structural features found in the parallelepipeds were performed. Figure 11.4 shows a reference spectrum of

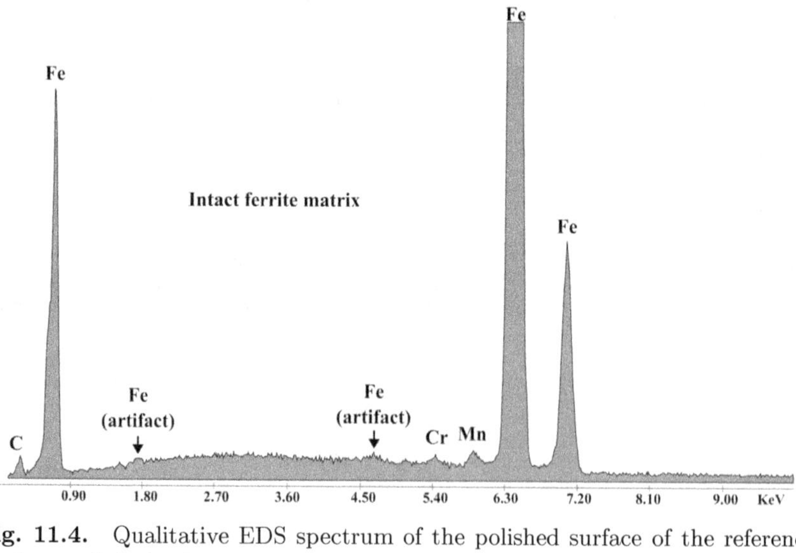

Fig. 11.4. Qualitative EDS spectrum of the polished surface of the reference ferrite parallelepiped.

the homogeneous intact ferrite, while Fig. 11.5 reports two representative X-ray spectra of materials on the external damage mark and inside the microcavities. The ferrite matrix, as expected, is composed of iron with minor amounts of carbon, chromium and manganese. The intact ferrite spectrum also shows two artefact peaks with energies matching the Si Kα1,2 and the La Lα1,2 peaks, a consequence of the high Fe count rate.

These peaks, as well as those characteristic of Fe, C, Cr and Mn, can also be found in the spectra of the external damaged zone and of the microcavities in the bar. In addition, the spectra of the damage mark, see Fig. 11.2(c), show peaks of elements not present in the intact ferrite, such as oxygen, sodium, aluminium, sulfur, chlorine, potassium, calcium and copper, as well as a more pronounced carbon peak, see Fig. 11.5. This enrichment in carbon is particularly evident in the spectra of the formless material filling the microcavities, Figs. 11.3(c) and 11.3(b), which is also characterized by lower amounts of oxygen and sulfur with respect to the damage mark, see Fig. 11.5.

The EDS data of the anomalous materials on the surface and inside the bar are shown in Tables 11.3 and 11.4, together with the average composition and the standard deviation (σ) of 12 external and internal analyses of the intact ferrite. The lower amount of data for the amorphous material inside the microcavities, see Table 11.4, is due to the difficulties inherent in the analyses of chaotic material inside the microcavities. Indeed, several microanalyses of this material had to be discarded due to low number of X-rays detected, which are expressed as counts s^{-1}.

The composition of the intact ferrite is fairly homogeneous. Besides iron (Fe = 98.5 wt.%), small amounts of carbon (C $\leq$ 0.1 wt.%), chromium (Cr = 0.4 wt.%) and manganese (Mn = 1.0 wt.%) were detected in the intact ferrite. The σ values indicate a high overall precision with relative uncertainties that increase from iron (0.1%) to carbon (21%), with decreasing contents of the analysed element, see Table 11.3.

The low-brightness material in the external damage zone, see Fig. 11.2(c), shows variable compositions characterized by lower Fe (83.4–90.0 wt.%) and Mn (0.5–0.9 wt.%) contents and higher C and Cr values (0.2–1.4 and 0.4–2.0 wt.%, respectively) than the intact ferrite. In addition, the damage mark shows remarkable and variable

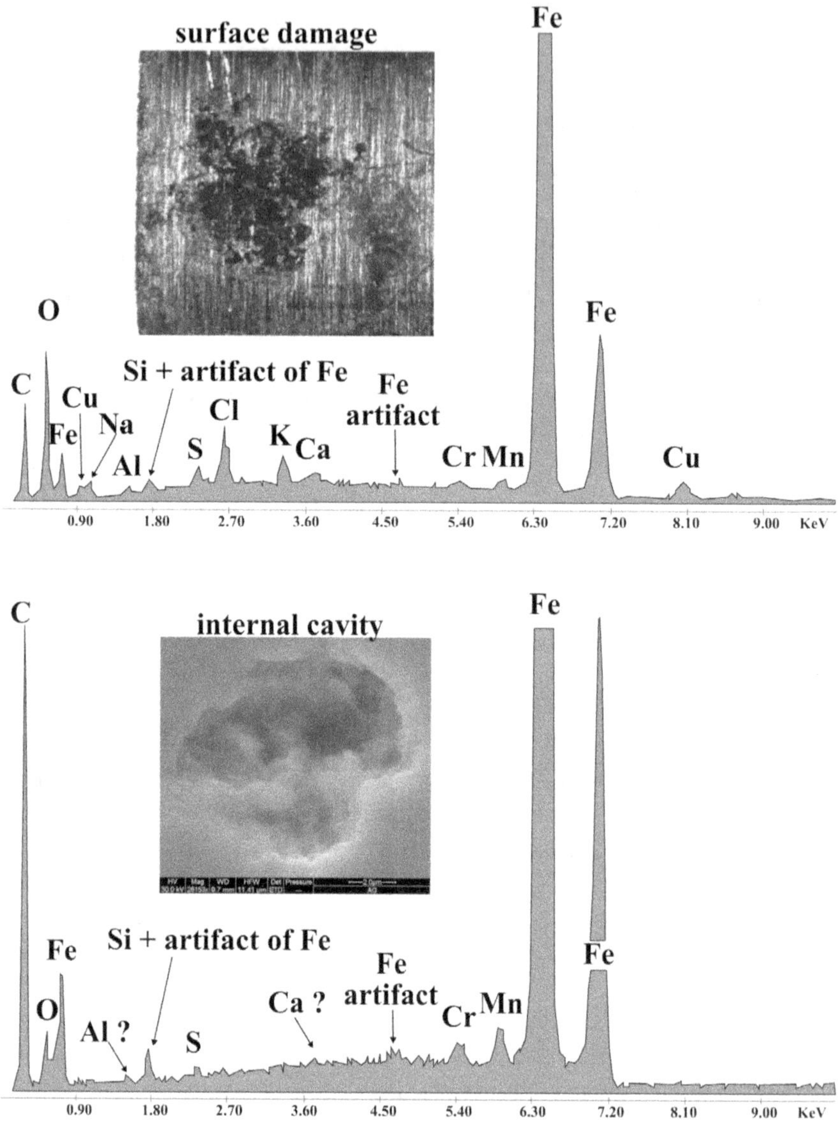

Fig. 11.5. Qualitative EDS spectra of the surface damage mark and of the amorphous material inside an internal cavity of the ferrite bar. The two spectra were obtained at the same operating conditions and have the same vertical counting scale.

Table 11.3. ESEM-EDS semi-quantitative data (in weight %) of the amorphous material in the external damage zone.

	Intact Ferrite		External Damage Material												
Spot	Average	σ (12)	ED1	ED2	ED3	ED4	ED5	ED6	ED7	ED8	ED9	ED10	ED11	ED12	ED13
C	0.08	0.02	0.40	0.24	0.23	0.33	0.21	0.37	0.26	0.57	0.75	0.67	1.37	1.10	1.05
O			4.27	5.86	5.84	6.13	6.91	6.85	6.58	4.14	3.31	2.19	2.81	3.26	3.88
Na			0.08	0.12	0.13	0.19	0.10	0.19	0.14	0.24	0.21	0.22	0.07	0.10	0.18
Mg											0.20	0.01			0.10
Al									0.14			0.10			
Si				0.09	0.21	0.07	0.17	0.16	0.13	0.12	0.29	0.09	0.36	0.23	0.30
S			0.48	0.41	0.46	0.59	0.51	n.m.	n.m.	0.57	0.54	0.49	0.25	0.25	0.36
Cl			1.76	2.80	2.86	2.75	2.74	2.71	2.70	2.20	1.98	1.75	1.62	1.53	1.48
K			0.99	1.17	1.38	1.09	1.29	1.27	1.20	1.07	1.07	1.06	1.16	0.99	0.87
Ca			0.46								0.99	0.48	0.25	0.22	0.22
Cr	0.43	0.05	0.41	1.84	1.82	0.97	2.04	1.79	1.69	0.71	0.71	0.62	0.79	0.48	0.61
Mn	0.97	0.08	0.66	0.78	0.70	0.60	0.65	0.57	0.71	0.66	0.64	0.91	0.48	0.55	0.57
Fe	98.52	0.11	89.74	84.81	84.32	83.43	83.62	83.45	83.54	86.85	87.21	89.82	89.75	90.05	89.05
Cu			0.75	1.88	2.05	3.84	1.76	2.66	2.92	2.86	2.11	1.59	1.09	1.23	1.33

Notes: The average composition and standard deviation (σ) of 12 analyses of the intact ferrite (including both external and internal zones) are also reported. n.m. = identified but not measured. Blank cells indicate that the element was not detected.

Table 11.4. ESEM-EDS semi-quantitative data (in weight %) of the amorphous material inside the microcavities.

	Intact Ferrite		Internal Damage Material						
Spot	Average	σ (12)	MC1	MC2	MC3	MC4	MC5	MC6	MC7
C	0.08	0.02	0.25	0.67	0.49	0.90	1.40	0.65	2.93
O				0.82	0.88	1.20	0.33	0.40	6.58
Al							0.08		0.35
Si			0.11	0.15	0.21	0.16	0.12	0.07	
S							0.08	0.14	
Ca								0.18	
Cr	0.43	0.05	0.47	0.51	0.47	0.68	0.68	0.91	1.00
Mn	0.97	0.08	1.02	1.14	1.12	1.10	1.25	1.10	1.43
Fe	98.52	0.11	98.15	96.71	96.83	95.95	96.05	96.56	94.29

Notes: The average composition and standard deviation (σ) of 12 analyses of the intact ferrite (including both external and internal zones) are also reported. Blank cells indicate that the element was not detected.

amounts of elements foreign to the matrix of the intact ferrite, such as oxygen (O = 2.2–6.9 wt.%), chlorine (Cl = 1.5–2.9 wt.%), potassium (K = 0.9–1.4 wt.%) and copper (Cu = 0.7–3.8 wt.%), as well as smaller concentrations of sodium, silicon and sulfur ($\leq$0.6 wt.%). Small amounts ($\leq$1.0 wt.%) of magnesium, aluminium and calcium are found only occasionally, see Table 11.3.

The X-ray microanalyses of the jagged material inside the microcavities of the bar interiors show a lower number of foreign elements, see Table 11.4. Na, Mg, Cl, K and Cu are apparently missing and the concentrations of Al, Si, S and Ca are vanishingly small ($\leq$0.2 wt.%). The contents of iron and oxygen measured in these small cavities (94.3–98.2 and 0.0–1.2 wt.%, respectively) are intermediate between those of the damage mark and the intact ferrite but closer to the ferrite composition. Astonishingly enough, the analyses of the microcavity material systematically show higher C, Cr and Mn concentrations than the intact ferrite, see Table 11.4. The exsolution, or rather the remobilization in the solid state, of these elements during the ultrasound treatment appears to be rather improbable since the material within the microcavities should have been associated with components poor in these elements and rich in Fe with respect to the ferrite.

The structural characteristics of the damage mark and the material composing it, see Fig. 11.2, suggest that it is the result of violent reactions within subsurface zones, which possibly have led to the partial fusion of the ferrite and the sudden fracturing and expulsion of the superficial layers of the bar. This is supported by the presence of bright remnants of pristine ferrite surface between the cratered dark material, see Figs. 11.2(c) and 11.2(d), and by the limited depth ($\sim$20 μm) of the craters, see Figs. 11.2(e) and 11.2(f). In addition, the cracking of the cratered material, as in Fig. 11.2(e), suggests that the material generated by the reaction was liquid and interacted with the surrounding atmosphere cooling rapidly. Furthermore, parts of the fused material were most likely expulsed during the explosive phase as suggested by its minimal thickness in cross-sectional view, see Fig. 11.2(f). This is in agreement with the presence of characteristic ferrite peaks, those of Fe, Cr and Mn, in the EDS spectra of the damage mark, see Fig. 11.5, which suggest that the ESEM electron beam penetrated the entire thickness of the damaged material and was interacting primarily with the ferrite matrix, see Fig. 11.4 for reference spectrum.

The irregular morphology of the material inside the microcavities, Figs. 11.3(b) and 11.3(c), cannot be an artefact of the low resolution of the ESEM because, at the operating conditions, the used microscope, FEI Quanta 200, is capable of distinguishing objects with dimensions as small as 3 nm. It is therefore safe to assume that the material inside the microcavities is composed of a chaotic assemblage of submicron-sized, likely amorphous particles. The EDS spectra of the material in the microcavities, see Fig. 11.5, are inferred to be the result of the summation of X-rays originating from the materials within the microcavities, to a small degree, and those of the ferrite matrix underneath, to a large degree, just as for the damage mark. The second spectrum in Fig. 11.5 shows that the material inside the microcavities is predominantly composed of carbon.

All of this evidence indicates that the amount of material foreign to the ferrite is minimal. This is supported by the relatively high Fe contents in all the spot analyses of the external damage mark, see Table 11.3, which is in contrast with the low BSE brightness of the material, see Fig. 11.2(c). Indeed, while backscattered electrons come from relatively shallow levels of the analysed materials, the emitted X-rays derive from a bulb-shaped volume extending deeper into the sample. Even though the diameter of the ESEM electron beam is $\leq 0.5\ \mu$m, its penetration depth and the extent of its interaction with the sample depend on the voltage used and the nature of the material being analysed. For example, the Kanaya–Okayama formula [5], which takes into consideration the ESEM beam energy, 30 kV, and the physical properties of the analysed material, gives beam penetration depths of 3.2 μm and 9.9 μm for ferrite and amorphous carbon materials, respectively. This indicates that most of the measured X-rays come from the ferrite lattice below the damage mark and microcavity materials. Therefore, the material of the external damage mark should have a greater thickness than the materials in the microcavities, because it shows lower Fe contents. Using a simple mass balance equation that takes into consideration the composition of the underlying ferrite matrix, it has been estimated that during the ESEM-EDS analysis of the external damage mark and the internal microcavities, the electron beam interacted with materials containing 8.8–15.6% and 0.4–4.3% of mass foreign to the ferrite, respectively. The conclusion is that the elements which are primary constituents of the amorphous material are O, Cl, S, K and Cu. We shall do a brief note later about the nature of the Cu.

The large number of highly deformed cavities inside the sample containing the damage mark, see Fig. 11.3, strongly supports the occurrence of DST reactions. In fact, their maximum size, $\sim$10 μm, is consistent with the theoretical calculations of the space time deformation induced by cavitation, see Tables 11.1 and 11.2. According to these calculations, the power of the ultrasound applied to the bar, transferred energy density to the gas inside the internal cavities enough to guarantee their collapse and the triggering of the subsequent DST reactions, producing the presence of the chaotic amorphous materials detected. Following these evidences, it is quite straightforward to conclude that the external damage, characterized by cratering and elements foreign to the ferrite, is the product of DST reactions taking place within gas-filled subsurface microcavities, which typically form during metal casting.

On the other hand, the attempts by Ridolfi *et al.* to hypothesise some other processes that could produce the observed damage and chemical changes have had little success.

The reaction products inside the microcavities are most likely the result of spherically symmetric collapse. This collapse has triggered the production of elements such as C, Mn and Cr, which in the analysis have shown linear variation patterns with iron. In the external damage zones, the spherically symmetric collapse most likely ended with the catastrophic rupture of the cavity walls, thus producing the observed cratering. Therefore, the spherically symmetric collapse is followed by an explosive event characterized by shock wave velocities in excess of those produced by the spherically symmetric collapse itself, $4v_s$, where v_s is the velocity of sound in the iron. This is in agreement with the effects of a violation of the Lorentz invariance at microscopical level predicted by DST theory. As a result, the spherically symmetric collapse-triggered production of carbon on the surface of the bar was followed, or accompanied, by the microexplosion-triggered formation of elements foreign to the ferrite. All this appears to be in agreement with the observed nonlinear, curvilinear behaviour of carbon and chromium measured in materials within the external cratering, see the data from Tables 11.3 and 11.4. In addition, the convergence of the linear and curvilinear data distributions resulted in the analysis supports a two-stage process whereby the carbon produced during the symmetrically spherical collapse and iron already present in the matrix are involved in further DST reactions, via catastrophic explosions, producing elements foreign to the ferrite: O, S, Cl, K and Cu. Now, we like to briefly report, without discussion, the evidence about Cu obtained by Induced Neutron Activation Analysis (INAA). The Cu showed an 'inverse abundance' of its main isotopes with respect to their 'natural abundance', allowing us to say that the detected Cu is 'not natural', and this is another element about its production by DST reaction.

Moreover, the correspondence of the internal microcavities and the centre of the external damage zone along the entire length of the bar suggests that the violation of the Lorentz invariance during ultrasound treatment only takes place in sections of the bar where the energy per unit time, or better the energy density in the time, is sufficient to trigger the reactions. The evidence that inside the ferrite bar there is a lack of amorphous materials in cavities smaller than 5 μm, whereas they are present in cavities of $\sim$10 μm, brings us to the conclusion of what the optimal cavity size r for DST reactions is: 5 $\mu\text{m} < r \leq 10$ μm. This is in strong agreement with the results reported in Tables 11.1 and 11.2, where the predictions of DST theory are collected.

The element changes due to DST reactions induced by ultrasound can be summarized as follows:

The material in the microcavities shows a systematic increase in carbon (C), chromium (Cr) and manganese (Mn) and a decrease in iron (Fe), see Table 11.4.
The surface damage zone shows a systematic increase in carbon (C), oxygen (O), sulfur (S), chlorine (Cl), potassium (K), chromium (Cr) and copper (Cu) and decrease in manganese (Mn) and iron (Fe), see Table 11.3.

The elements foreign to the ferrite bar are as follows: O, oxygen; Cl, chlorine; K, potassium; and Cu, copper, which have been determined to be 'not natural' [4,6].

These results are in agreement with the theoretical model predicting the deformation of the space time and the violation of the local Lorentz invariance, and suggest that nuclear reactions characterized by 'not-radioactive transmutation' phenomena and the emission of neutron bursts accompanied by an absence of gamma emissions can occur following the application of mechanical pressure alone. These DST reactions in ferrite bars appear to take place in two distinct and successive steps: symmetrically spherical collapse and catastrophically explosive rupture. At equal applied pressures, these reactions appeared to depend on the size of the microcavities in the ferrite ($\leq$10 μm) since the amount of energy per unit time accumulated during the symmetrically spherical collapse increases with decreasing gas bubble size as described in the previous section.

In the previous section, we have found a correlation between the power setting and the bubble/cavity size. In light of the possible triggering of the DST reactions by the presence of microcavities in solids and cavitation bubbles in liquids, we perform a comparison of the power setting of the ultrasound machine and the density difference between the treated materials and air, the presence of which in both the cavitation bubbles and the ferrite microcavities is likely. Table 11.5 demonstrates that the values obtained by dividing the applied power by the air to condensed matter matrix, ferrite or water density ratio are very similar, 1.3×10^5 and 1.1×10^5, respectively. This suggests that the triggering of DST reactions by ultrasound treatment of materials with densities higher than that of iron will occur at power settings below the 19 W used in the experiments with steel and ferrite bars. This evidence paves the way, as we will see, to the search for DST reactions in higher density liquid materials, such as mercury.

In addition, there is an agreement between the distances travelled by the ultrasound waves in the two materials, iron and water, and the start of neutron emissions. By multiplying the speed of sound in iron, $v_{s(\text{ferrite-steel})} = 5800$ m/s, by the time interval between the start of the experiment and the emission of the first neutrons, 300 s, the distance travelled by the ultrasound waves was calculated as 1.7×10^6 m. However, looking to the cavitator sonotrode set up with respect to the ferrite bar, see Fig. 6.7, we have to take into consideration that the machine generates two sets of ultrasound wave-fronts

Table 11.5. Relationships between the power applied during the ultrasonic experiments and the air, ferrite and water mass to volume densities (D).

Matrix	D (Kg/m^3)	$\frac{D_{\text{air}}}{D_{\text{matrix}}}$	Power (W)	$\frac{\text{Power}}{D_{\text{air}}/D_{\text{matrix}}}$ (W)
Ferrite	7874	1.5×10^4	19	1.3×10^5
H_2O solution	997	1.2×10^3	130	1.1×10^5

Table 11.6. Densities of the microcavity-rich zone in the ferrite bar calculated using a maximum cavity diameter of 10 micron.

$D_{n/V}$ (m^{-3})	$D_{n/M}$ (kg^{-1})	$D_{v/V}$	$D_{m/M}$
5×10^{13}	6×10^{9}	3×10^{2}	4×10^{6}

Notes: n: number of cavities, v: cavity volume, m: cavity air mass, V: ferrite volume, M: ferrite mass; air mass inside cavity was calculated assuming the air density at standard conditions (1.2 kg/m^3); the density used for the ferrite matrix is 7874 kg/m^3.

of opposite direction, which means that the ferrite bar is crossed by the ultrasound two times, the total distance travelled within the ferrite is therefore doubled, 3.4×10^6 m. This value is very similar to the distance estimated for the experiments in water solutions, 3.6×10^6 m, calculated for one wave-front travelling with a sound speed, $v_{s(\text{water})} = 1480$ m/s, and a time interval between the start of the experiment and the first neutron emission of 2400 s.

The different energy densities of the materials, see Table 11.5, and thus the different amount of time before the start of the reactions, are a function of the differences in the matrix densities, liquid or solid, crossed by the ultrasound waves and that of the air in the similar-sized bubbles or cavities, respectively. Table 11.6 lists the estimated parameters for the microcavity-rich zone inside the ferrite bar, or rather maximum diameter and densities of the cavities in the ferrite. The listed densities represent maximum values as they were estimated using the maximum observed cavity size of 10^{-5} m. The ratio between the cavity and ferrite masses, D_m/M, also represents a maximum value as it was calculated assuming the maximum possible density of gas present inside the cavities during their collapse, using the density of the air at standard conditions, atmospheric pressure and room temperature. Table 11.6 is another useful tool to look for DST reactions in other materials, or in producing an iron bar having the right cavity size and the right cavity density in order to use them as 'fuel bars'.

Lastly, we remark that these microcavities have nothing to do with the Kirkendall voids, whose dimensions range in the interval 40–60 micron, as is well known in metallurgy.

At the end of this chapter we can state that the prediction, according to the DST theory of the microreactor size, 4–8 micron, has been fully verified by the discovery of microcavities in which if the size is below 5 micron, no reaction occurs, whereas if the size is within 10 micron, DST reactions take place, giving rise to the production of elements other than the matrix material. These cavities working as DST microreactors have been discovered by Ridolfi and are so named after him, 'Ridolfi cavities'.

References

[1] L. Crum and D. Gaitan. Frontier of nonlinear acoustic. *Proceedings of 12th International Symposium on Nonlinear Acoustic.* Elsevier Applied Science, New York, 1990.

[2] C. Brennen. *Cavitation and Bubble Dynamics.* Oxford University Press, 1995.

[3] F. Ridolfi *et al. Appl. Phys. A*, 114, 1233–1246, 2014. DOI: 10.1007/s0039-013-7876-z.

[4] F. Ridolfi *et al. J. Radioanal. Nucl. Chem.*, 304(2), 955–963, 2015. DOI: 10.1007/s10967-014-3341-5.

[5] K. Kanaya and S. Okayama. *J. Phys. D App. Phys.*, 5, 43, 1972. DOI: 10.1088/0022-3727/5/1/308.

[6] A. Rosada *et al. J. Adv. Phys. Spec. Sect. New Nucl. React.*, 5(1), 90, 2016. DOI: 10.1166/jap.2016.1243.

Chapter 12

Critical Energy Densities

We have seen in Section 11.1 that, according the DST theory, it is possible to use the interaction energy thresholds $E_{0\text{int}}$ for obtaining a phenomenological energy threshold E_{thres} triggering both the DST reactions and emissions. But in Section 7.2 we saw that it is the energy density which rules the DST emissions and the consequent DST reactions. Now we face the task to estimate for each interaction, when possible, the 'critical energy densities' both in the space and in the time. For doing this we will use the size of the Ridolfi cavities and the evaluation of the time interval for anisotropic interaction. In general we define critical energy density its evaluation at the energy threshold $E_{0\text{int}}$, or rather at the 'boundary' where the interaction energy parts the flat from the deformed space time. More in general from a theoretical and phenomenological point of view the spatial energy density via the use of the size of the Ridolfi cavities gives us the way to interpret the spatial homogeneity Pessa constant introduced in Section 4.3. We can now phenomenologically identify the Pessa constant with the size of the Ridolfi cavity. We do not emphasize this identification since in Chapter 9 we separated the need of homogeneity from the physical units realizing it.

12.1 Critical Densities in the Space

We use the spherical approximation, as in Section 11.1, in using the space volume given by space size ranging in the interval 4–8 micron, theoretical, or 5–10 micron, phenomenological. So the critical volume we use is $V_c = 5.2\ 10^{-10}\,\mathrm{cm}^3$. Now we are entitled to estimate the critical energy densities at the leptonic and hadronic thresholds, considering these two interactions as a 'couple' within the recursive pattern of the interactions, as described in Section 2.5 and recalled at the beginnings of the Section 7.2. Therefore at the energy threshold the leptonic critical energy density is given by the estimation of $\mathrm{D_{lep}} = 1.5\ 10^{11}\,\mathrm{GeV/cm^3}$, and the hadronic critical energy density is given by the value $\mathrm{D_{had}} = 7.0\ 10^{11}\,\mathrm{GeV/cm^3}$. As we note these, nearly close, energy densities come from nothing else than space time geometrical properties including the interaction energy. Therefore we warn and strongly stress that they have nothing to do with the results obtained within a flat space environment as, for example, the Nuclear Density Functional theory (DFT), which is build on theorems showing that all many-body correlations, like say in nuclei, bring to the existence of universal energy functionals. It is quite evident that the many-body correlations live in the Euclidean, at most Minkowskian, geometry. In any case these values are well far from the standard nuclear density expressed in energy units, which is roughly $\mathrm{D_{nucl}} = 2\ 10^{-39}\,\mathrm{GeV/cm^3}$.

Now we look for critical energy densities in the space for electromagnetic (e.m.) and gravitational interactions. We use as reference value the energy density in the cosmic vacuum as from anisotropy of the Cosmic Microwave Background (CMB) radiation: $\mathrm{D_{CMB}} = 5\ 10^3\,\mathrm{eV/cm^3}$. We perform the evaluation using again the spherical approximation for the critical volume V_c. The results with respect to $E_{0\mathrm{e.m.}}$ and $E_{0\mathrm{grav}}$, the other couple in metrics recursive pattern, are collected in Table 12.1 where the size ranges in the interval 1–10 micron.

We can see from the underscored values in Table 12.1 that the match with $\mathrm{D_{CMB}}$ value occurs for $\mathrm{D_{e.m.}}$ for size R ranging in the interval 5–6 micron, the lower limit of the Ridolfi cavities, and for $\mathrm{D_{grav}}$ for size R ranging in the interval 9–10 micron, the upper limit of the Ridolfi cavities. Obviously we cannot claim this match as a proof, but we consider it a clue, intriguing enough. We will be back about CMB and $\mathrm{D_{CMB}}$ in the last Part V of the book, where we consider the cosmological consequences of DST theory once endowed with the asymmetry besides the anisotropy.

Table 12.1. Values for e.m. and gravitational critical densities $D_{e.m.}$, D_{grav} for size R in the interval 1–10 micron, the underscored values match the D_{CMB} value.

Size R *micron*	Spherical V_c cm^3	$D_{e.m.}$ eV/cm^3	D_{grav} eV/cm^3
1	4.19×10^{-12}	1.07×10^{6}	4.82×10^{6}
2	3.35×10^{-11}	1.34×10^{5}	6.03×10^{5}
3	1.13×10^{-10}	3.98×10^{4}	1.79×10^{5}
4	2.68×10^{-10}	1.68×10^{4}	7.53×10^{4}
5	5.24×10^{-10}	8.59×10^{3}	3.86×10^{4}
6	9.05×10^{-10}	4.97×10^{3}	2.23×10^{4}
7	1.44×10^{-9}	3.13×10^{3}	1.41×10^{4}
8	2.14×10^{-9}	2.10×10^{3}	9.42×10^{3}
9	3.05×10^{-9}	1.47×10^{3}	6.62×10^{3}
10	4.19×10^{-9}	1.07×10^{3}	4.82×10^{3}

12.2 Critical Time Interval

We already used the concept of characteristic time evolution in Section 11.1, in order to handle the DST reactions, and now we recall here some main features. Borrowing methods and notation from the general theory of relativity, we can state that, for a particle at rest:

$$\frac{dt_{\text{had}}}{dt_{\text{e.m.}}} = \frac{1}{\sqrt{g_{0,\text{had}}}}. \tag{12.1}$$

Since for the deformed metric it is $g_{0,\text{had}} = b^2_{0,\text{had}}(\text{E})$, we get

$$\frac{dt_{\text{had}}}{dt_{\text{e.m.}}} = \begin{cases} 1, & 0 \leq E \leq E_{0,\text{had}} = 367,5 \pm 0.4\,\text{GeV} \\ E_{0,\text{had}}/E, & E_{0,\text{had}} < E \end{cases}. \tag{12.2}$$

Starting from Eq. (12.2) and using the 'hadronic energy speed', or speed in the fifth dimension W_{had}, as in Section 11.1, we can define the critical time in the following way:

$$dt_{\text{had}} = E_{0had}/W_{\text{had}}. \tag{12.3}$$

The Eq. (12.3) is defined at the neighbourhood of the hadronic metric $E_{0\,\text{had}}$, but as the 'critical time interval' independent from the energy E supplied via e.m. interaction. Using $E_{0\,\text{had}} = 367.5\,\text{GeV}$ and $W_{\text{had}} = 4.8\ 10^{21}$ GeV/sec we get

$$dt_{\text{had}} = 76.5 \cdot 10^{-21}\,\text{sec}. \tag{12.4}$$

With an abuse of notation we can use W_{had} to estimate dt_{lep} in the so called 'flat region' of the two metrics, hadronic and leptonic, between the energy values $E_{0\,\text{lep}} \le E \le E_{0\,\text{had}}$ of the metric couple:

$$dt_{\text{lep}} = E_{0\text{lep}}/W_{\text{had}}. \tag{12.5}$$

Using $E_{0\,\text{lep}} = 80.6\,\text{GeV}$ and as above $W_{\text{had}} = 4.8\ 10^{21}$ GeV/sec we get

$$dt_{\text{lep}} = 16.8 \cdot 10^{-21}\text{sec}. \tag{12.6}$$

12.3 Critical Densities in the Time

We simply note that in both cases, hadronic and leptonic, the 'critical time interval' we estimated are about three order of magnitude above the so called 'Yukawa time' which is of the order of 10^{-23} sec. The meaning of the critical time intervals is that at the neighbourhood of the boundaries of the above quoted metric 'flat region' DST reactions and connected DST emission can occur within these intervals. Moreover they give an evaluation of the time interval within which the spatial energy density has to be reached at microscopical level. Regarding the DST emissions, which we examined in Chapters 6 and 7, these critical time intervals give the tool to interpret the fact that such emissions occur in particle bursts giving rise to some hardness in detecting them, and therefore forcing to use many kinds of detector performing measurements both differential and integral, like those ones in Section 6.3, in order to get a 'double cross check' of the emission phenomenon.

We can handle the concept of the critical time interval for the metric couple hadronic – leptonic, whereas we cannot do the same with the couple e.m. – gravitational due to the difficulty in establishing an analogous of the Eq. (12.2) for these interactions. Therefore we only can say that the main energy density in time in order to obtain DST reactions in condensed matter is

$$W = 4.8 \cdot 10^{21}\,\text{GeV/sec.} \tag{12.7}$$

Together with the space densities we can trigger and possibly control the process of energy density inside the condensed matter. This process has been called 'energy densification' by F. Pistella, who introduced it in examining any kind of DST reaction which can occur in condensed matter producing both isotopic changes and new elements production, together with the emission of nuclear particles without concomitant gamma rays emission [1] and [2].

12.4 Pressure and Power

If we look for DST reactions, both at small and large dimensions of energy, space and time, it is useful to define the estimated energy densities in practical units.

To this aim we can consider that an energy density in the space, within a spatial volume, well corresponds to a pressure, not in vectorial sense or rather unidimensional in direction, but in the sense of the Pascal Principle. In the meanwhile we can consider an energy density in a time interval well corresponding to a power, see Section 11.1. The union of suitable pressure and power can be used to give rise, trigger and possibly control the 'energy densification' in condensed matter regardless the practical way to supply power and pressure, since in any case they are produced by devices and supplied to matter ever by electromagnetic interaction. Now we report the estimated densities in units of Pascal and watt.

$$D_{\text{had}} = 1.12 \cdot 10^{8}\text{Joule/m}^3 = 112\,\text{MegaPascal.} \tag{12.8}$$

$$D_{\text{lep}} = 2.5 \cdot 10^{7}\text{Joule/m}^3 = 25\,\text{MegaPascal.} \tag{12.9}$$

$$W = 7.6 \cdot 10^{11}\text{Joule/sec} = 760\,\text{GigaWatt.} \tag{12.10}$$

By the way, as we saw the D_{had} is useful, at at least at microscopical level, to produce bursts of DST neutron, meanwhile D_{lep} is useful, at at least at microscopical level, to produce bursts of DST alpha.

At the end of any process of energy densification if as microreactors are used microbubbles in the liquids, like water solutions, and

Table 12.2. Relationships between the power applied during the ultrasonic experiments and the air, ferrite and water mass to volume densities (D).

Matrix	D (kg/m^3)	$\frac{D_{air}}{D_{matrix}}$	Power (W)	$\frac{Power}{D_{air}/D_{matrix}}$ (W)
Ferrite	7874	1.5×10^4	19	1.3×10^5
H_2O solution	997	1.2×10^3	130	1.1×10^5

Table 12.3. Densities of the microcavity-rich zone in the ferrite bar calculated using a maximum pore diameter of 10 micron.

$D_{n/V}$ (m^{-3})	$D_{n/M}$ (kg^{-1})	$D_{v/V}$	$D_{m/M}$
5×10^{13}	6×10^9	3×10^2	4×10^6

Notes: n: number of cavities, v: cavity volume, m: cavity mass, V: ferrite volume, M: ferrite mass.
air mass inside cavity was calculated assuming the air density at standard conditions ($1.2\,kg/m^3$); the density used for the ferrite matrix is $7874\,Kg/m^3$.

Ridolfi cavities in the solids, like iron or steel, another problem must be faced, or rather the number of these microreactors useful to produce DST reaction at macroscopical level, see Tables 12.2 and 12.3. This means that the number of microreactors working in energy densification conditions must engage in the DST process a number of atoms, and their nuclei, comparable with the Avogadro number. Therefore we have resumed here these three issues needed to build up any device producing in stable way DST reactions by using the indications from Eqs. (12.8)–(12.10) together with Tables 12.2 and 12.3. Even if there are two more issues. The first one is to avoid the reaction becoming divergent and uncontrolled, but this can be obtained remaining in the 'flat region' between the two E_{0int} of the metrics couple, which unfortunately is a very narrow condition but practicable. The second one is the 'angle law' which we will treat in the Part IV of the book.

References

[1] F. Pistella *et al.* *Symmetry*, 15(8), 1507, 2023. DOI: 10.3390/sym 15081507.

[2] S. Bellucci *et al.* *Symmetry*, 13(4), 674, 2021. DOI: 10.3390/sym 13040607.

Part IV
Energy Asymmetry

Chapter 13

Angle Law: Magnetic Coil

Generally speaking, the Clausius postulate, applied to electrical systems, can be seen as a first form of the law of conservation of their total electric charge. In the same way, the deformed space time (DST) reactions obey the law of conservation of the electric charge of the affected atomic nuclei. Therefore, for instance, the 'genesis' of alpha particles and neutrons is not the production and result of nuclear radiation, but fragments coming from nucleosynthesis and nucleolysis.

Clausius postulate can also be considered as the application of the second law of thermodynamics to electric charge systems, as atomic nuclei are as well atomic electrons including conduction electrons. Namely, if there is no kinematic decoupling between the charges and the field they generate, also in presence of an electric charge gradient in the field reference system, electromagnetic emissions are absent, because there is no motion of the charges. Moreover ever according to Clausius postulate if there is no kinematic decoupling between the charges and an external steady field, no electric charge gradient is possible at all, therefore no difference of electrical potential or rather no non-zero voltage is detectable. In summary, according to the Clausius postulate, in a static electrical system even in presence of an external steady field the absence of any non-zero voltage is the signature of validity of local Lorentz invariance (LLI). In simple words, when there are static electrical charges and fields, or rather without kinematic decoupling, it means that no voltage difference occurs and all this is a signature of LLI validity.

Now the use of the Clausius postulate is at the very foundation in the issue of testing the LLI in an electromagnetic system using the electrical charges inside a steady field. Again, if there is no kinematic decoupling between the charges and an external steady field, an electric charge gradient in the field reference system, detectable as a non-zero voltage, can be a signature of LLI violation.

Following this premise we recall the proposal put forward by Umberto Bartocci, in the last decade of the XX century, under the inspiration he got from the work of Stefan Marinov. Bartocci [1] proposed a new electromagnetic experiment aimed at testing the LLI and able to provide direct evidence for its violation. The results obtained in the first experiment, carried out in June 1998, were essentially aimed at providing new upper limits on the LLI breakdown parameter and they admit, as the most natural interpretation, the fact that local Lorentz invariance is in fact broken or violated [2,3]. The experiment was repeated, at L'Aquila Italy, in 1999 in a different laboratory, 100 Km far from the previous one, using a completely new and improved apparatus. The analysis of this second experiment confirmed the positive evidence of the previous one [4]–[6].

Let us consider a steady current I circulating in a closed loop C, and a charge q subjected to the uniform magnetic field produced by C.

If both q and C are at rest in the same reference frame K in which electromagnetism can be described by standard Maxwell theory, then no force F acts on q due to C. In fact, the expression of this force is:

$$F = q\left(-\nabla\Phi - \frac{\partial A}{\partial t} + V \times B\right). \tag{13.1}$$

where Φ and A are respectively the electric and magnetic potential associated to C, B = curl A the corresponding magnetic field, and V the velocity of q in K and, in the aforesaid conditions, $\Phi = 0$, $\frac{\partial A}{\partial t} = 0, V = 0$. Notice that, as a matter of fact, the equation $\Phi = 0$ is just an additional hypothesis in classical electrodynamics, the above mentioned Clausius postulate, which holds for 'ideal' one-dimensional circuits. For real ones, the validity of this equation has been questioned both on an experimental basis and on a theoretical one, but in any case without relevance for the experiments performed.

Let us now introduce a reference frame K' moving with some uniform velocity $\mathbf{v}$ with respect to K. Then, by means of the Principle of Relativity, one can assert that there is no force acting on q, if both q and C are at rest in this new frame K'. In order to compute the required force without principle of relativity, we can use the following Eq. (13.2), where of course V becomes equal to v, and we suppose, according Clausius Postulate, that $\Phi = 0$, an hypothesis which makes all the difference with the relativistic interpretation of Maxwell electrodynamics, precisely that difference we have just to look for. We call x, y, z, t the space-time coordinates in K, and suppose that:

- the velocity v is equal to the constant velocity $(v, 0, 0)$;
- the "moving" circuit is parametrized by the equations $x = rcos\theta + vt, y = rsin\theta, z = 0$;
- the charge q is placed in the point $P = (vt, y, 0)$.

Then straightforward calculations yield the expression which is approximated up to terms of second order in $\beta = v/c$:

$$F = -\frac{\mu_0 qIv}{4\pi r}\left\{\int_0^{2\pi}\frac{sin^2\theta - \frac{y}{r}sin\theta}{\left[1 - \frac{2\frac{y}{r}}{1+(\frac{y}{r})^2}sin\theta\right]^{3/2}}d\theta\right\}e_y, \tag{13.2}$$

with e_y being the unit vector in the y-direction.

Instead of a single charge, let us consider a tiny conductor R of length $L < r$, lying on the (x, y)-plane, with an end in the center O of the loop.

Then, in order to find the voltage $V(s)$ across the conductor, one has simply to integrate the field corresponding to Eq. (13.2) with respect to y, thus getting

$$V(s) = \frac{\mu_0 Iv}{4\pi}\int_0^s [1 + u^2]^{-3/2}\left\{\int_0^{2\pi}\frac{usin\theta - sin^2\theta}{\left[1 - \frac{2u}{1+u^2}sin\theta\right]^{3/2}}d\theta\right\}du,$$

$$\approx 10^{-7} Ivi(s) \tag{13.3}$$

where we put $s = L/r < 1$, and $i(s)$ is the following double integral

$$i(s) = \int_0^s [1+u^2]^{-3/2} \left\{ \int_0^{2\pi} \frac{usin\theta - sin^2\theta}{\left[1 - \frac{2u}{1+u^2} sin\theta\right]^{3/2}} d\theta \right\} du, \qquad (13.4)$$

which can be numerically evaluated for different values of s.

Of course, the existence of the above voltage is strictly related to the relative motion of the conductor and its conduction electrons with respect to the reference frame K, or, better, with respect to the magnetic field $\mathbf{B}$. Therefore, a measurement of a $V \neq 0$ for a system of a coil and a conductor, at rest in the laboratory frame, constitutes a direct evidence of the violation of local Lorentz invariance. Assuming the kinematic point of view, that we used up to now, such a violation can be parametrized in terms of a speed v trough Eq. (13.3), and one of its possible interpretations is assuming that actually the magnetic field produced by the coil 'is not at rest with it', but subdued to a kind of 'kinematical decoupling' without violation of Clausius postulate. Assuming again the kinematic point of view, it is possible to say that the system of a coil and a conductor is at rest with respect to some reference frame, just moving with respect to Earth with speed v, which drags it. It is worth stressing that such an effect does not depend on the magnitude of the speed v as compared to the light speed c, and one may be able to observe it even for very small v by suitably increasing I. One would have therefore, in this case, a violation of LLI at low speeds, contrarily to what is generally believed.

Moreover, it is expected, on the basis of the deformed space time (DST) theory, that this effect critically depends on the orientation of the plane where the system lies. This is reflected in the structure of the experimental apparatus devised to look for it, as we shall see.

The test is based on the possibility of detecting a non-zero Lorentz force between the magnetic field B generated by a stationary current I circulating in a closed loop C, and a charge q, in the hypothesis that both q and C are at rest in the same inertial reference frame. Such a force is zero, according to the standard relativistic electrodynamics.

The experimental setup was devised in order to put new upper limits on the violation of LLI, by means of such an entirely new class of electromagnetic experiments, and also to test possible anisotropic and asymmetric effects in such limits.

The experimental device used in both experiments is schematically sketched in Figs. 13.1 and 13.2, together with the electric schemes of the apparatus and the measurement device. It consisted of a Helmholtz coil C and a Cu conductor R placed inside it on a plane orthogonal to the C axis. The conductor R was connected in series to a capacitor C, and a voltmeter was connected in parallel to the capacitor, so to measure the voltage due to a possible gradient of charge across R.

The conductor could change its orientation in the coil plane. Moreover, the whole system of the RC circuit and the coil could turn so letting its plane coincide with one of the coordinate planes.

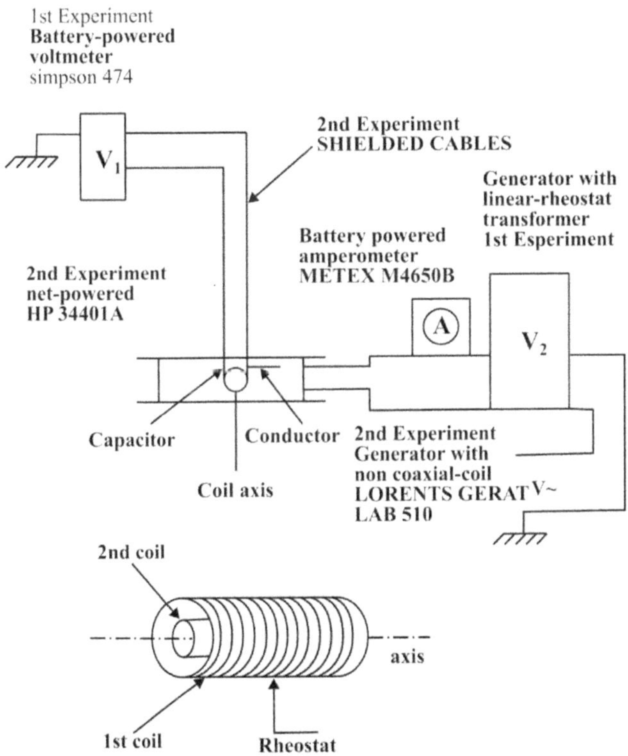

Fig. 13.1. Schematic front view of the Helmholtz coil and the capacitor-resistor probe.

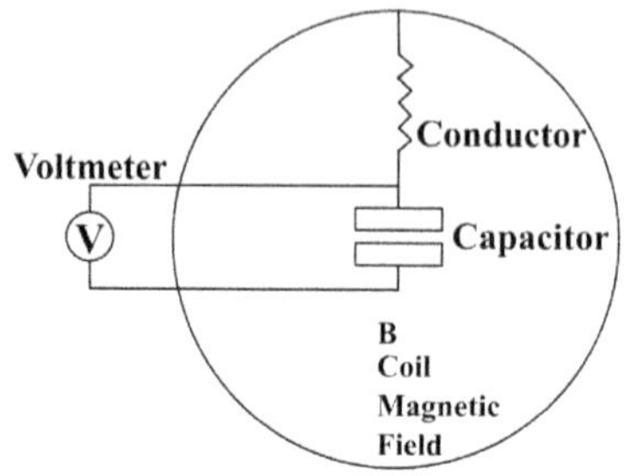

Fig. 13.2. Schematic view of the experimental setup.

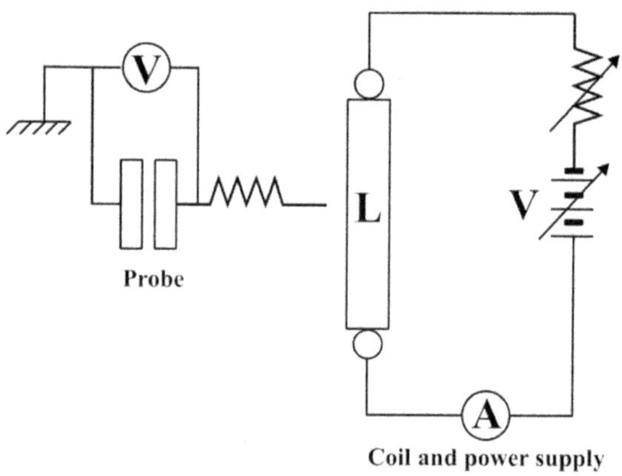

Fig. 13.3. Sketch of the electric circuits of the capacitor-resistor probe and the Helmholtz coil with power supply.

The center of the geometrical coordinate system coincided with the center of the coil. The coordinate system was chosen as follows: the (x, y) plane tangent to the Earth surface, with the y-axis directed as the local Earth magnetic field $\mathbf{B}_T$; the z-axis directed as the outgoing normal to the Earth surface, and the x-axis directed so that the coordinate system is left-handed. The conductor orientation in the coil plane was parametrized in terms of an angle α, ranging from 0 to 2π. The rotation of α was chosen clockwise in the plain coil with respect to an observer oriented along the coordinate axis orthogonal to the coil plane. The first orientation of R corresponding to the angle $\alpha = 0$ was along the negative direction of the z-axis in the case of the two vertical canonical planes, see Fig. 13.4 for the (x, z) plane, and along the negative direction of the y-axis in the case of the horizontal plane. A steady-state current I circulating in the coil produced a constant magnetic field B in which the RC circuit was embedded. The circuit and the coil were mutually at rest in the laboratory frame.

The measurements performed with the system lying on the plane (y, z) gave values of V compatible with the instrument zero, see Fig. 13.4(c). Indeed, in such cases the statistical tests of correlation showed that each of the points outside the zero-voltage band is uncorrelated with the preceding and the subsequent point either, and the whole set of points was shown to be uncorrelated ($R^2 < 30\%$). Let us recall that each point is the average of five measurements, taken at the same angle.

The measurements in the planes (x, y) and (x, z), shown instead that the points outside the zero band are statistically correlated ($R^2 > 80\%$), and so they represent valid candidates for a nonzero signal, see respectively Figs. 13.4(a), 13.4(b), 13.5(b) and 13.6.

For the plane (x, z) the LLI violation is in a broad angular interval. A polynomial interpolating curve for the points related to (x, z) plane is shown in Fig. 13.6. Such an interpolating procedure was essentially aimed at finding the angle $\alpha_{\max}$ corresponding to the maximum value of V, $V^{xz}_{\max} = (36.0 \pm 1.0)\ \mu V$. The value found was $\alpha_{\max} = 3.757\,\text{rad}$. The knowledge of $\alpha_{\max}$ is needed in order to determine the value of the anisotropic LLI violation parameter. On the contrary for the (x, y) plane we see a narrower angular interval corresponding to an angle $\alpha = \pi/4\,\text{rad}$.

Now we face for the first time a true anisotropic and even asymmetric situation, which suggests that the LLI violation has a geometrical asymmetric violation.

For comparison, the results of the first experiment are given in Table 13.1.

The corresponding voltage value will be referred to as a 'peak value'. In absence of peaks, one calculates the average of all the measured values outside the instrumental zero the 'level value'.

In the second experiment, a signal in the form of a peak signal was analogously found again in the planes (x, y) and (x, z). About (x, z) plane for $B = B_1 = (5.14 \pm 0.01)\,\text{mT}$, the average peak value was in excellent agreement with the result of the former experiment: $V'^{xz}_{\max} = 35.4 \pm 0.1\,\mu V$. The signal was again highly anisotropic, and its behaviour with α is the same as depicted in Fig. 13.6. However,

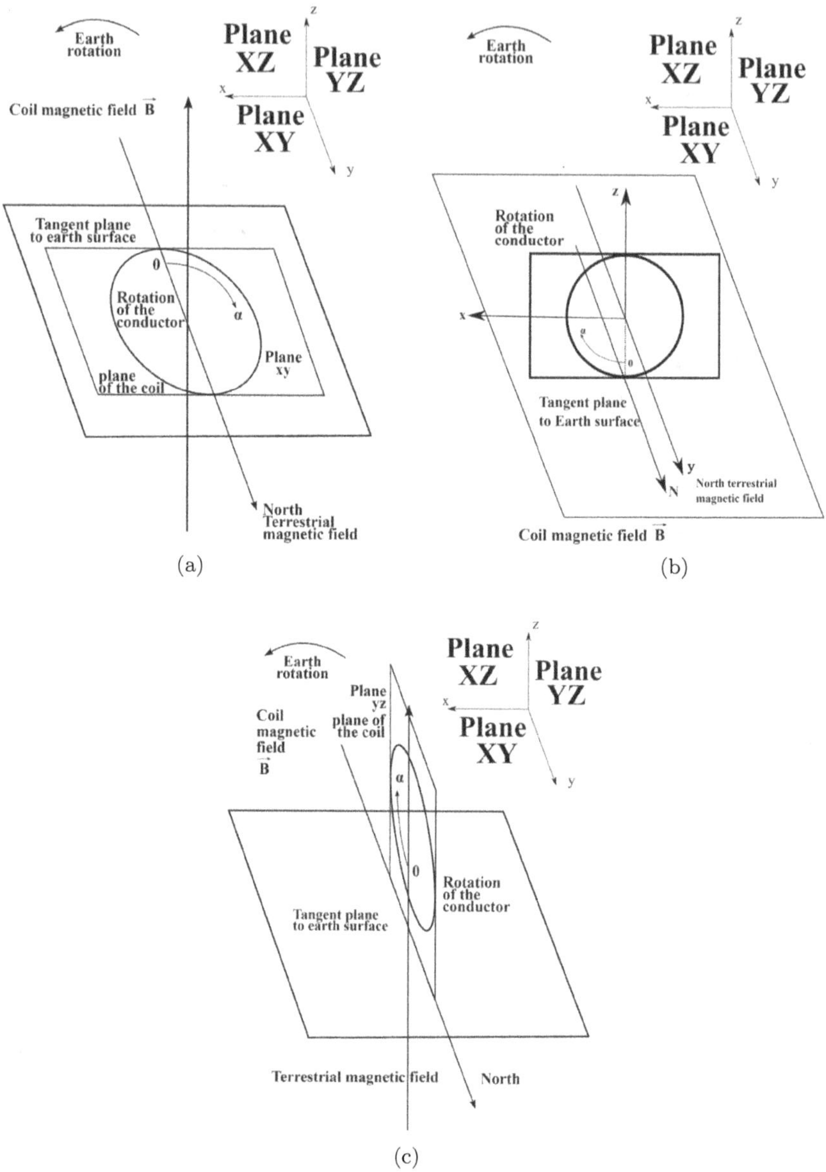

Fig. 13.4. (a) Schematic view of the orientation of the apparatus and the related magnetic fields in the xy plane. (b) Schematic view of the orientation of the apparatus and the related magnetic fields in the xz plane. (c) Schematic view of the orientation of the apparatus and the related magnetic fields in the yz plane.

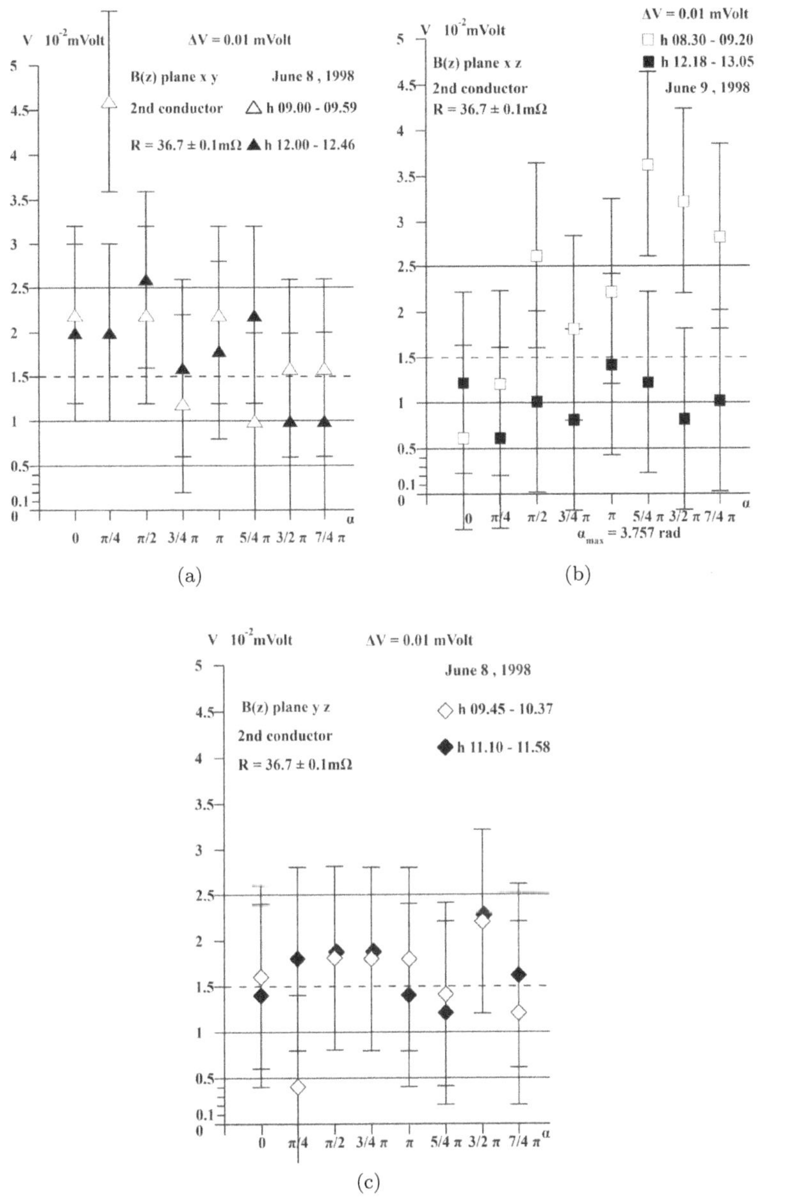

Fig. 13.5. (a) Results of the measurements with the apparatus in the xy plane. $\Delta V = 0.01\,mV$ is the error on the voltage V across the capacitor. (b) Results of the measurements with the apparatus in the xz plane. $\Delta V = 0.01\,mV$ is the error on the voltage V across the capacitor. (c) Results of the measurements with the apparatus in the yz plane. $\Delta V = 0.01\,mV$ is the error on the voltage V across the capacitor.

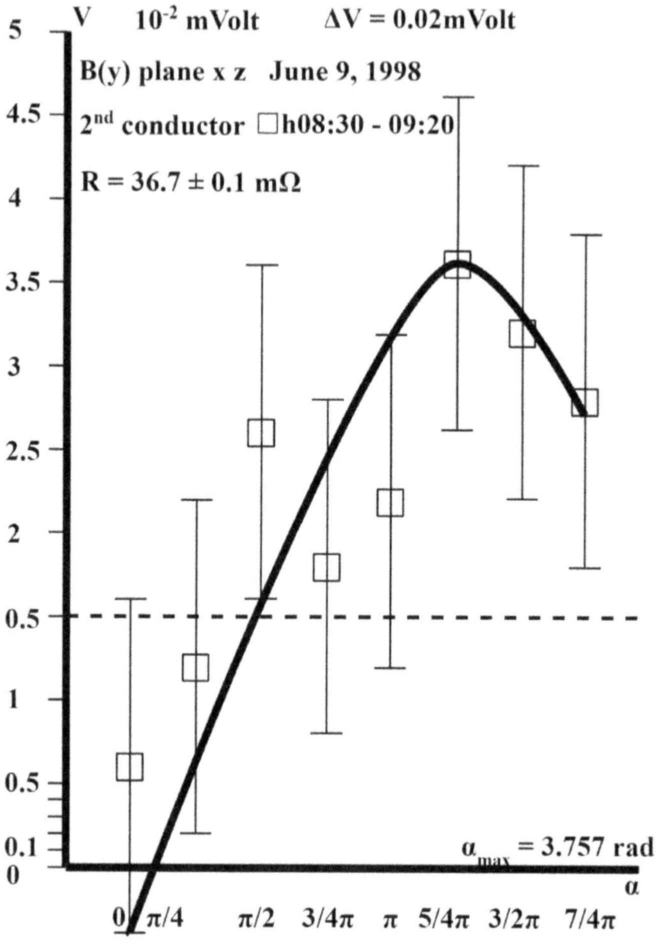

Fig. 13.6. Curve interpolating the data obtained with the apparatus in the (x, z) plane, showing the angle of maximum signal $\alpha_{\max} = 3.757$ rad.

now possible signal candidates were found also in the plane (y, z); this evidence is also a consequence of the higher sensitivity of the multimeter, improved by two orders of magnitude. In that plane, there was no dependence on α, and therefore no spatial anisotropy. On the contrary, a time anisotropy was found in the (x, y) plane, since the measurements taken a.m. gave values within the instrument zero. The average level values found for $B = B_1$ were $V'^{xy} = (30.7 \pm 0.1)\,\mu V$ and $V'^{yz} = (26.6 \pm 0.1)\,\mu V$. The measurements taken with

Table 13.1. Results of the 1998 experiment for the measured V. Symbol: V_{am} = average of a.m. measurements.

Plane	B = 3.65 mT V(μV)
xy	$V_{am} = 17.4 \pm 0.3$
yz	$V_{am} = 15.9 \pm 0.2$
xz	36.0 ± 1.0
	$V_0 = (15.0 \pm 10.0)\mu V$

the halved value of the coil magnetic field, B = B_2 = (2.58±0.01) mT, gave similar results, with voltage values V'^{xz} max = $(41.8 \pm 0.1)\,\mu$V, $V'^{xy} = (34.4 \pm 0.1)\,\mu$V and $V'^{yz} = (30.6 \pm 0.1)\,\mu$V. Not only these values were not halved with respect to those obtained for B = B_1, as expected in the case of a linear relation between V and I, like that derived via the Lorentz force, but surprisingly enough they were slightly higher! Moreover, a check was made by reversing the coil current. No change in the sign of V occurred. This allows to conclude that the observed effect is independent of the direction of the current and depends apparently in the 'wrong way' on its magnitude.

The results obtained in the second experiment are summarized in Table 13.2.

In the case of the (x, y) plane, no value of <V> is reported because the value of V_{am} was too narrow in angular distribution unlike what occurs in the (x, z) plane.

In analogy with the 1998 experiment, the adopted criterion in order to accept a voltage peak in a given plane was the presence of at least three points statistically correlated with maximum value of voltage at least 2σ from the other uncorrelated points.

An attempt at interpreting the above experimental results can be given in terms of the DST metrics. To this aim, let us first discuss in more details the implications of the spatial anisotropy of the observed effect. To this purpose we define the following quantities:

$$D \equiv V_{am}^{xz} - V_{am}^{xy}; \tag{13.5}$$

$$\Delta \equiv \langle V_{am}^{xz} \rangle - V_{am}^{xy}. \tag{13.6}$$

The values of D, Δ are given in Table 13.3. Notice that such values increase with decreasing B. This fact will be discussed later on.

Table 13.2. Results of the 1999 experiment for the measured V. Symbols: $V_{am(pm)}$ = average of a.m. (p.m.) measurements; $\langle V \rangle = (V_{am} + V_{pm})/2$.

Plane	$V_{am}(\mu V)$	$V_{pm}(\mu V)$	$<V>(\mu V)$	isotropy	
	$B = B_1 = 5.14\,mT$			Space	Time
xy	17.7 ± 1.6	30.7 ± 1.5		Yes	No
yz	24.1 ± 2.4	29.1 ± 2.1	26.5 ± 1.8	Yes	Yes
xz	35.3 ± 0.5	37.0 ± 0.5	36.2 ± 0.7	No	Yes
	$B = B_2 = 2.58\,mT$				
xy	16.8 ± 1.6	34.4 ± 1.5		Yes	No
yz	27.0 ± 1.7	34.1 ± 1.4	30.6 ± 1.7	Yes	Yes
xz	43.8 ± 0.3	39.7 ± 0.3	41.8 ± 0.7	No	Yes
V_0	16.5 ± 1.5			Yes	

The averages $\langle D \rangle$, $\langle \Delta \rangle$ are therefore

$$\langle D \rangle = (21.1 \pm 0.9)\mu V; \quad \langle \Delta \rangle = (20.7 \pm 1.0)\mu V. \tag{13.7}$$

Let us attempt an interpretation of these results.

Both experiments show a peak signal in the (x, z) and (x, y) planes. Such an effect could, at a first sight, be attributed to a gravitational breakdown of the LLI. However, the results obtained in the first and second experiment, which show a narrow 'angular' signal in the plane (x, y), suggest that the observed effect is both gravitational and electromagnetic. If so, then it is possible to conclude that the difference between the measured voltages in the two planes, namely, the quantities $\langle D \rangle$, $\langle \Delta \rangle$, provides a measure of the amount of the gravitational contribution to the measured voltage versus the electromagnetic one.

This is indeed supported by the agreement of the values in Eq. (13.7) with the threshold energy for the gravitational metric $E_{0,grav}$.

On the other hand, let us evaluate the differences in the voltages measured in the three planes. According to the second experiment with $B = B_1$, see Table 13.3, the transitions of the voltage values corresponding to the LLI violation occur when coming through different canonical planes.

Table 13.3. Comparison of the results for the first and second coil experiments.

B (mT)	D(μV)	Δ(μV)
$B_1 = 5.14$ (I exp.)	17.7	18.5
$B = 3.65$ (II exp.)	18.6	18.6
$B_1 = 2.58$ (I exp.)	27.0	25.0

First, from the space anisotropy in the plane (x, z) ($\langle V^{xz}\rangle = (36.2 \pm 0.7)\mu V$) to the time anisotropy in the plane (x, y) ($\langle V^{xy}_{pm}\rangle = (30.7 \pm 1.5)\mu V$).

Second from the time anisotropy in the plane (x, y) to the space-time isotropy in the plane (y, z), ($\langle V^{yz}\rangle = (26.5 \pm 1.8)\mu V$). Both of them do occur for steps $\langle V\rangle \simeq 4.85\,\mu V$:

$$\langle V^{xy}_{pm}\rangle - V^{xy}_{pm} = 5.5\mu V; \quad V^{xy}_{pm} - \langle V^{yz}\rangle = 4.1\mu V. \tag{13.8}$$

This does just agree with the value of the threshold energy for the electromagnetic metric $E_{0,\text{em}}$. In general, we expect that the transitions can occur for multiples, by relative integers, of $\Delta V \sim 5\,\mu V$. This occurrence can envisage or better clues a sort of 'quantization of metric variation', but we do not like to push further this possibility.

The fact that both electromagnetic and gravitational metrics must be taken into account in order to interpret the results of this type of experiments, namely either interaction is responsible of the observed LLI violation, can be also suggested by the value of the instrumental zero V_0 in both experiments, $V_0 \propto (E_{0,\text{grav}} - E_{0,\text{e.m.}})$, the proportionality constant being obviously the electron charge.

The relative difference in sign between the gravitational and electromagnetic contributions to V_0 is essentially due to the opposite behaviour of the two corresponding metrics. In fact, the gravitational metric is over-Minkowskian, and reaches the limit of Minkowskian metric for decreasing values of E with $E > E_{0,\text{grav}}$, whereas the electromagnetic metric is sub-Minkowskian and thus attains the Minkowskian limit for increasing values of the energy $E < E_{0,\text{e.m.}}$. Therefore the two metrics become Minkowskian for energy variations of opposite sign. In this context, it is worth noticing two intriguing features of V_0. On one side, the energy value corresponding to V_0 lies inside the energy interval ($E_{0\text{el}}$, $E_{0\text{grav}}$) where either metric is exactly

Minkowskian; moreover, such a value coincides with the amplitude of that interval of Minkowskian behaviour. This circumstance seems to provide also an explanation of the apparent independence of the instrumental zero from the measuring device.

We can therefore conclude that, even if the first experiment left open the possibility of a mere gravitational explanation of the observed effect, the results of the second one are strongly in favour also of an inescapable electromagnetic contribution to the violation of the LLI.

Let us attempt also a qualitative explanation of the strange fact, see Table 13.2, that there are voltage values slightly higher for the halved value B_2 of the magnetic field in the second experiment.

It is indeed $V_{am}^{xy}(B_1) \simeq V_{am}^{xy}(B_2)$ but $V_{pm}^{xy}(B_2) > V_{pm}^{xy}(B_1)$; $\langle V_{pm}^{yz}(B_1)\rangle > \langle V_{pm}^{yz}(B_1)\rangle$; $\langle V^{xz}(B_2)\rangle > \langle V^{xz}(B_1)\rangle$.

In particular, one has

$$\begin{aligned} d^{xy} &\equiv V_{pm}^{xy}(B_2) - V_{pm}^{xy}(B_1) = 3.7\mu V; \\ d^{yz} &\equiv \langle V^{yz}(B_2)\rangle - \langle V^{yz}(B_1)\rangle = 4.1\mu V; \\ d^{xz} &\equiv \langle V^{xz}(B_2)\rangle - \langle V^{xz}(B_1)\rangle = 5.6\mu V. \end{aligned} \tag{13.9}$$

And therefore

$$\langle d\rangle = 4.5\mu V, \tag{13.10}$$

again in agreement with the value of $E_{0,\text{e.m.}}$. This effect can be therefore interpreted as an electromagnetic one. When the voltage values are within the instrumental zero V_0, we are in a region where both metrics are Minkowskian and there is then no difference between the two values corresponding to B_1 and B_2. For $V > V_0$, decreasing the value of the magnetic field implies to decrease the energy of the system. According to the sub-Minkowskian behaviour of the electromagnetic metric, this implies a stronger departure from the Minkowski metric, or rather an enhanced effect in the violation of the Lorentz invariance which manifests itself in a higher value of the voltage, contrarily to what expected on classical electrodynamic arguments based on the Lorentz force. This point is further enforced by the values of the quantities D and Δ, as introduced in Eqs. (13.5) and (13.6), which increase with decreasing B, see Table 13.3.

The problem of the LLI violation adimensional parameter δ and how it is very difficult to detect such an effect by means of an experiment of the Michelson–Morley type, or rather by interferometric apparatuses, is widely examined in [2,3]. The conclusion of this exam is that the kinematic velocity v, decoupling the reference frames, has a value of about 6 cm/sec, which is too little with respect say the Earth rotation 4.6 10^4 cm/sec, Earth revolution 3 10^6 cm/sec or worse the Galaxy drift velocity 6 10^7 cm/sec. This evidence compelled us to consider this Lorentz violation due to other cause, since effectively this value of v 'maybe' is within the measure error for the above quoted velocities. Looking to Figs. 13.4 and 13.5 it is possible to state that within the deformed space time the Lorentz violation is no longer kinematic but geometric. This geometric violation has the properties to be anisotropic and asymmetric or rather 'directional' in the sense that can identify violation angles as the angles α obtained from Fig. 13.5(a) and 13.5(b) with Fig. 13.6: $\alpha = \pi/4$ rad

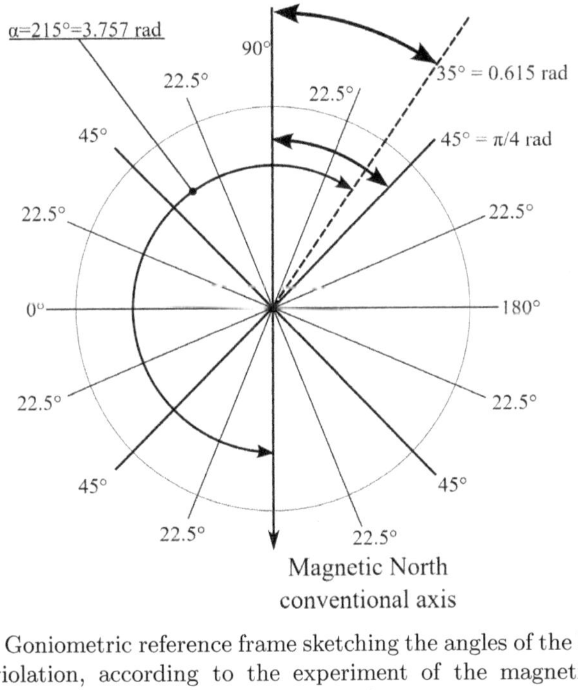

Fig. 13.7. Goniometric reference frame sketching the angles of the local Lorentz invariance violation, according to the experiment of the magnetic coil, α = 3.757 rad and $\pi/4$ rad. Here the angles $\alpha = 215°$ and its reduction to first quadrant or rather $\alpha - 180° = 35°$ are shown.

and $\alpha = 3.757\,\text{rad}$. These angles give a situation where there is a 'fundamental asymmetry' underlying the natural phenomena when the space time deformation occurs. In Fig. 13.7 we resume in a goniometric reference frame these angles which from now on can be called the 'angle law' of the asymmetry Lorentz violation in deformed space time. The connection of this asymmetry with the energy as fifth coordinate is the quest we treat in the following of this book.

References

[1] U. Bartocci and M. Mamone Capria. *Found. Phys.*, 21, 787, 1991.

[2] U. Bartocci *et al. Found. Phys. Lett.*, 14, 51, 2001.

[3] F. Cardone and R. Mignani. *Phys. Essays*, 13, 643, 2000.

[4] F. Cardone and R. Mignani. *Gravitation, Electromagnetism and Cosmology: Toward a new Synthesis*, p. 165 Ed. by K. Rudnicki, Apeiron, Montreal 2001, Proc. Int. Conf. On 'Redshifts and Relativistic Universe', Cesena, Italy, Sept. 17–20, 1999.

[5] F. Cardone and R. Mignani. Proc. Int. Conf. On 'Physical Interpretation of Relativity Theory VII', Imperial College London Sept. 15–18, 2000. *Late Papers and Supplementary Papers*, p. 40 Ed. by M. Duffy.

[6] R. Scrimaglio *et al. Found. Phys.*, 36, 263, 2006.

Chapter 14

Angle Law: Shadow of Light

14.1 Inverse Compton Effect

It is well known that the scattering of thermal photons from energetic electrons results in high energy photons, in the gamma range, and this is called the inverse Compton scattering. In the 1980s of XX century had been performed measurements of these inverse Compton scattering gamma photons at CERN using the electron beam of the Large Electron Positron collider (LEP). The so produced photons in the scattering off thermal radiation, of the LEP vacuum pipeline, at temperature $T = 300°K$, up to 6 GeV dominate over the 'bremsstrahlung' radiation photons owing to residual gas in the pipeline and this fact is mainly due to low gas pressure of about 5×10^{-10} Torr. The inverse Compton scattering in this case can be schematically resumed in the sketch of Fig. 14.1. The results of the measurements are shown in Fig. 14.2, where are reported also the interpolations according both the hypotheses of the zero point radiation together with the stochastic radiation, or rather the radiation in the vacuum. These hypotheses need to introduce a cut-off energy E_c to avoid an ultraviolet divergence in the corresponding modified Planck distribution. The stochastic model of the zero point radiation increases the density of the 'incident' photons or 'target' photons but on one side it can give upper limit on zero point radiation, on the other side this result is at the price of losing the Lorentz invariance (LLI) since in quantum electro dynamics (QED) the radiation in the vacuum has energy density strictly zero because of the preservation of the LLI. The evaluation of E_c gives rise to an upper limit of the energy range of the LLI violation. According to the data reported in Fig. 14.2 the mean value for this cut-off energy at the temperature of 300°K is $< E_c >= 0.042\,\text{eV}$, which is about three order of magnitude above the $E_{0,\text{e.m.}}$ [1].

On the other hand the results depicted in Fig. 14.2 must be read in the inverse way according to the phenomenon sketched in Fig. 14.1. It means that in Fig. 14.2, the data at low energies, which are the produced photons, refer to higher energy incident photons of the thermal zero point-stochastic radiation. The good agreement with the zero

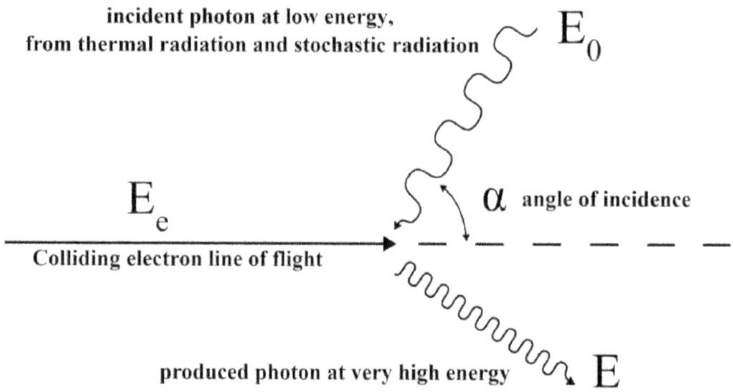

Fig. 14.1. Inverse Compton scattering: notation.

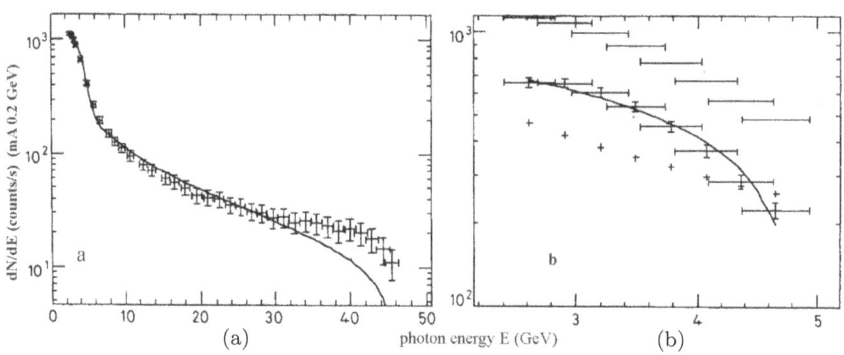

Fig. 14.2. (a) Fit of the spectrum for $E \geq 2\,\text{GeV}$; the low-energy cut has been introduced according to the recommendations of the detector builders, because of the performance of the apparatus. (b) Fit of the low-energy part of the spectrum. The points with only x-error bars are the data, the crosses are the bremsstrahlung fit, and the fitted points are the Compton obtained subtracting the bremsstrahlung fit from the data.

point-stochastic Planck distribution is evident in Fig. 14.2(b). On the contrary the data of the high energies produced photons, which come from low energy incident photons, show discrepancy with the zero point-stochastic Planck distribution, discrepancy increasing as the energy increases, or rather as the incident photon energy decreases. This evidence tell us that the hypotheses of zero-point radiation even improved with stochastic radiation cannot fully match the data. This situation, together with the fact that the use of energy cut-off E_c produces LLI violation, brings us to two possibilities. Either there is 'something else' among the photons at low energies, or there is 'something else' in the photons regardless the energy. Following this last possibility within the framework of deformed space time (DST) theory, where LLI violation is naturally embedded, it was looked for what is this 'something else' in the photon, in the sense that the results of Fig. 14.2(a) can even be interpreted as a sort of anomalous behaviour of the photons as particles. It seems that the particle photon should have an 'extension' working like a 'cross section' of a particle. In this sense after looking at the collision between photon and electron in order to test the above possibility it is worth to explore the collision between photon and photon, but in such conditions where no effect can happen in a physical situation preserving LLI. Therefore any anomalous behaviour detected must be compared with the features of DST theory, thus becoming another of her test.

14.2 Crossed Photon Beams

The results of the inverse Compton scattering measurements suggest the clue that anomalous effects can be observed also in different experimental situations involving colliding photon systems like in interference experiments. After the study of the 'collision' between electron and photon the evidence of an anomalous photon system behaviour can be searched in 'collision' between photons and the related anomalous 'photon-photon cross section'. This occurrence was observed indeed in experiments using orthogonal crossing photon beams.

These interference experiments were carried out by A. Ranfagni, at the National Research Council Institute of Applied Physics 'N. Carrara' (IFAC-CNR) in Florence Italy, with microwaves emitted by horn antennas, see the experiment layout in Fig. 14.3, [2, 3], and the other with infrared CO_2 laser beams, see the experimental scheme in Fig. 14.4, by R. Meucci at the National Research Council Institute of Applied Optics (INOA-CNR) in Florence Italy [4]. Let us summarize the results obtained.

The main result of the IFAC experiment consists in an unexpected transfer of modulation from one beam to the other, which cannot be accounted for by a simple interference effect. This confirms

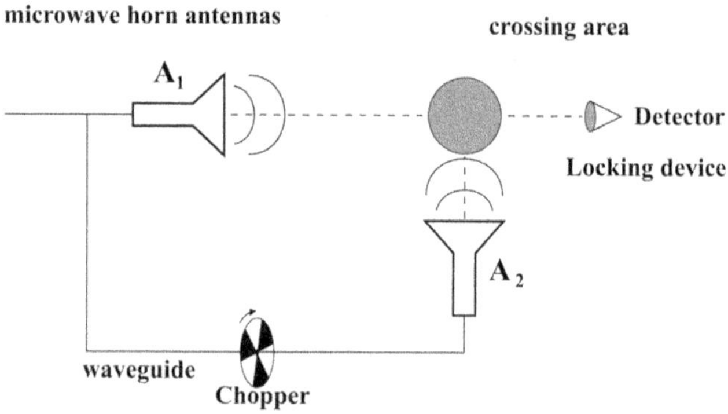

Fig. 14.3. Layout of the crossed-beam experiment in the microwave range using two horn antennas A_1 and A_2 *supplied by same e.m. wave by waveguide and* producing same orthogonal e.m. waves.

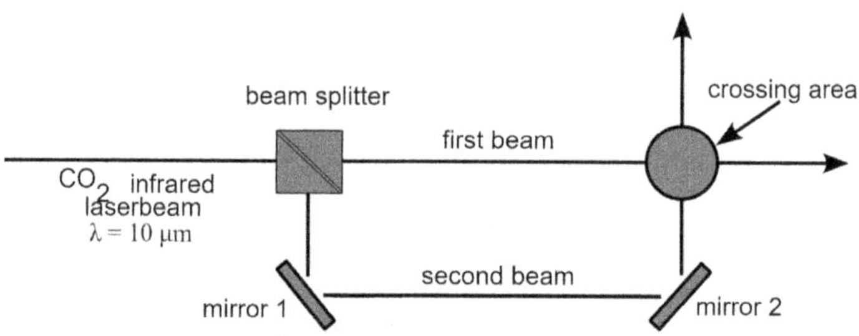

Fig. 14.4. Schematic view of the crossed-beam experiment in the infrared range, exploiting a CO_2 laser emitting at 10.6 μm on the fundamental TEM_{00} Gaussian mode. The laser beam is split in two orthogonal beams, beam 1 and beam 2, by means of a beam splitter. By using two flat mirrors the two beams are directed to the crossing area within the near field of the Gaussian mode, estimated at 1.5 m from the out-coupler mirror of the laser cavity. Beam 2 is periodically interrupted by means of a chopper whose frequency is the reference frequency in a lock-in amplifier connected to the detector.

the presence of an anomalous behaviour in photon systems, in the microwave range too.

In the optical experiment carried out at INOA-CNR, the wavelength of the used infrared laser beams was 10,600 nm. The optimum alignment which can be achieved with lasers and the laser beam confinement make this optical set-up especially suitable for investigating the anomalous behaviour of the photon systems. The measurements were carried out for a relatively long lapse of time, that is 12 min. This allowed one to perform a statistical test on the averaged results. The signal statistics provided a significant variation in the mean values obtained with or without beam crossing. Hence the chance to have two identical statistics was rejected with a sufficient level of confidence, see [4]. This allowed one to perform a statistical test on the averaged results. Moreover, by the analysis of INOA experimental data, it was estimated the displacement of the crossed beam signal with respect to the single beam signal. It resulted that the actual shift of the crossed beam signal with respect to the single beam signal is $2.08 \pm 0.13\,\mu$V, well below the threshold for electromagnetic metric $E_{0,\mathrm{e.m.}}$ and therefore within the deformed region of the e.m. metric which we remind in Fig. 14.5. Note that the laser experiment together with the microwave experiment show that the observed phenomenon does depend neither on the infrared wavelength, nor on the coherence properties of the light. Therefore one can conclude that the

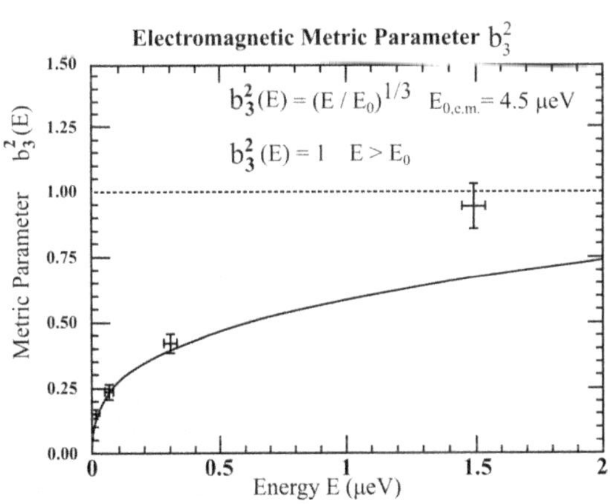

Fig. 14.5. Fit of the electromagnetic metric parameter $b_3^2 = b^2(E)$ vs. the energy of the output signal.

crossing photon-beam experiments give preliminary support to the evidence for an anomalous interference effect under the space and energy constraints obtained by the DST theory.

14.3 Double Slit Like Experiment

The following double slit like experiment we describe has noble roots in physics since proposed originally by L. de Broglie in a simplified layout and arrangement [5].

The experiments performed were optical ones, in the infrared range, of the double-slit type. They were essentially aimed at searching for a possible anomalous photon behaviour, at variance with the predictions of classical or quantum electrodynamics, and therefore directly related to Lorentz invariance (LLI) violation. Let us briefly report the main features and results of the three experiments, carried out at L'Aquila University [6–8].

The employed apparatus is schematically depicted in Fig. 14.6, it consisted of a Plexiglas box with wooden base and lid. The box was thoroughly screened from those frequencies susceptible of affecting the measurements and it contained two identical infrared (IR) light emission diodes (LED), as incoherent sources of light, and three

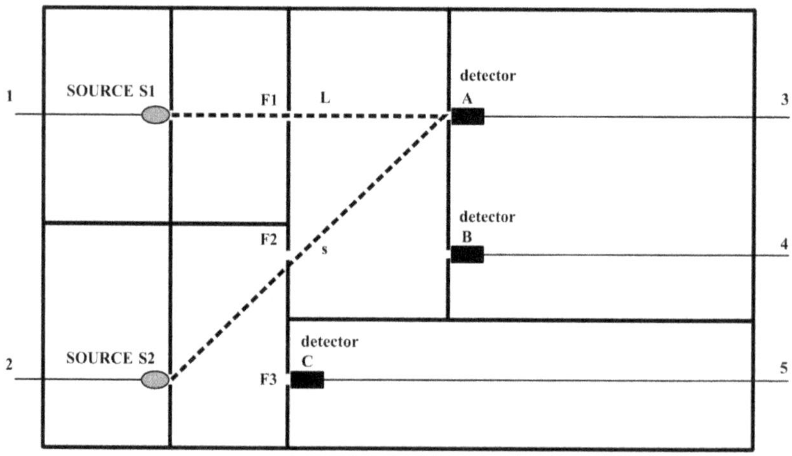

Fig. 14.6. Layout of the experimental apparatus used in the double-slit like experiment.

identical photodiodes or phototransistors, as detectors labelled in Fig. 14.6 with the letter (A, B, C). The two sources S1, S2 were placed in front of a screen with three circular apertures, or slits, on it labeled in Fig. 14.6 with the letter F1, F2, F3. The apertures F1 and F3 were lined up with the two LEDs A and C respectively, so that each IR beam propagated perpendicularly through each of them.

The geometry of this equipment was designed so that no photon could pass through aperture F2 on the screen. In this connection, let us note that the dotted line S in Fig. 14.6 is a mere geometrical one, and does not represent any physical trajectory of photons emitted by the source S2, since the aperture F2 was well outside the emission cone of S2. The wavelength of the two photon sources was $\lambda = 8.5 \times 10^{-5}$ cm, the slits were circular apertures, with a diameter of 0.5 cm, much larger than λ. Now, it is worth to remark again that this used wavelength of 850 nm is one order of magnitude lower than that one used in the laser crossed-beam experiment, which was of 10,600 nm. This was done in order to differentiate once more the range of energies explored. The whole system worked therefore in absence of single-slit Fresnel diffraction. However, the Fraunhofer diffraction was still present, and its effects have been taken into account in the background measurement.

Detector C was fixed in front of the source S2; detectors A and B were placed on a common vertical, movable panel frame, as we can see in Fig. 14.6. This latter feature allowed to study the space dependence of any anomalous effect, at variance with the cross beam experiments and the inverse Compton scattering measurements. Let us now highlight the role played by the three detectors. Detector C destroyed the eigenstates of the photons emitted by S2. Detector B ensured that no photon passed through the aperture F2. Finally, detector A measured the photon signal from the source S1. In summary, detectors B and C played a controlling role and ensured that no spurious and instrumental effects could be mistaken for the anomalous effect which had to be revealed on detector A. The design of the box and the measurement procedure were conceived so that detector A was not influenced by the source S2 according to the known and usually accepted laws of physics governing electromagnetic phenomena: classical or quantum electrodynamics. In other words, with regards to detector A, all went as if the source S2 was not there at all or as if it was always kept turned off.

In essence, the experiment just consisted in the measurement of the signal of detector A, aligned with the source S1, in two different states of source lighting. Precisely, a single measurement on detector A consisted of two steps:

1. Sampling measurement of the signal on A with source S1 switched on and source S2 off.
2. Sampling measurement of the signal on A with both sources S1 and S2 on.

As already stressed, due to the geometry of the apparatus, no difference in signal on A between these two source states ought to be observed, according to either classical or quantum electrodynamics. If $A(S1_i\ S2_k)$ ($i, k = \text{on}, \text{off}$) denotes the value of the signal on A when source S1 is in the lighting state i and S2 in the state k, a possible nonzero difference $\Delta A = A(\text{S1 “on” S2 “off”}) - A(\text{S1 “on” S2 “on”})$ in the signal measured by A when source S2 was off or on, and the signal in B was strictly null, has to be considered evidence for the searched anomalous effect.

The outcomes of the initial experiments were positive, namely the differences ΔA between the measured signals on detector A in the two conditions were different from zero and below the threshold value of energy for the violation of local Lorentz invariance (LLI) well within the framework of the deformed space time (DST) theory. In particular, ΔA ranged from $2.2 \pm 0.4\,\mu\text{V}$ to $2.3 \pm 0.5\,\mu\text{V}$, values well below the threshold $E_{0,\text{e.m.}} = 4.5\,\mu\text{V}$, therefore well inside the deformed part of the electromagnetic energy metric, see Fig. 14.5, and also in good agreement with the result of the measurements in laser cross beams experiment which gave for its analogous ΔA the value of $2.08 \pm 0.13\,\mu\text{V}$, as reported in the previous paragraph. Moreover, such an anomalous effect was observed within a distance of at most 4 cm from the sources, thus confirming the existence of spatial threshold like in the waveguide experiment, see Section 1.1. At a first glance this result can be considered such an effect due to the consequence of a sort of ‘hidden’ interference.

The purpose of the next set of experiments was to corroborate the results of the previous ones. The experimental set-up was essentially the same. For instance, the dimensions of the apparatus, and the relevant quantities, like photon wavelength and aperture diameter, were identical to those of the first group of experiments. The main difference with respect to the equipment of the first group of experiments was in a right-to-left inversion along the bigger axe of the box, and in the three used detectors, which were not photodiodes but phototransistors, of the type with a convergent lens. In this way, it was possible to study how the phenomenon changes under a spatial parity inversion and for a different type of detector. We want to point out that in this second set of experiments the time procedure to sample the signals on the detectors was different from that used in the first set. Indeed it has been realized that the sampling time procedure was apparently crucial in order to observe the anomalous interference effect.

The results of this second set confirmed those of the first one. The value of the difference measured on detector A was $(0.008 \pm 0.003)\,\mu$V, which is consistent, within the error, with the difference ΔA $= 2.3\,\mu$V measured in the first experiment, provided that the unlike efficiencies of the phototransistors with respect to those of the photodiodes are taken into account. It is possible to define the relative geometrical efficiency η_g of the phototransistor, with respect to the photodiode, as the ratio of their respective sensitive areas, and their relative time efficiency η_t as the ratio of their respective detection times. Then, we can define the relative total efficiency η_T of the phototransistor with respect to the photodiode as the product $\eta_T = \eta_g\, \eta_t$, from the values of η_g and η_t in this case, one gets $\eta_T = 0.0015$, see Tables 14.1 and 14.2.

Table 14.1. Parameters of the detector response (from the Data Sheet).

Experiment	Detector	Sensitive area (mm^2)	Detection time (μs)
1°	Photodiode	5.244	90
2°	Phototransistor	0.140	5

Table 14.2. Relative efficiencies of the phototransistor detector employed in the second set of experiments.

$\eta_{\rm g}$	$\eta_{\rm t}$	$\eta_{\rm T}$
0.0267	0.0556	0.0015

Therefore, it was reasonable to foresee that the value of the expected phenomenon in the second set of experiments had to be given by the product of the total relative efficiency times the value measured in the first set, or rather, $\eta_{\mathrm{T}}\ (2.3 \pm 0.5)\,\mu\mathrm{V} = 0.004 \pm 0.001\,\mu\mathrm{V}$. This result is in agreement with the second set experimental result, therefore both in agreement with the laser cross beam experiment and both within the deformed part of e.m. metric, see again Fig. 14.5. Again we see in this way how the double cross check method is applied.

The consistency between the results of the first two sets of experiments shows apparently that the effect is not affected by the parity of the equipment and by the type of detector used, at least for photodiodes and phototransistors. Let us notice that one was compelled to use two different sampling time procedures for the two different types of detecting devices in order to make the effect evident. It turned out that there was apparently a sort of unavoidable bond between detector and sampling-time procedure, to be taken into account in order to reveal the effect.

The third set of experiments was planned and carried out in order to shed some light on this issue and to obtain a further evidence of the searched effect. In order to test the apparent bond between detectors and sampling time procedures, the experiment was carried out by means of the box with photodiodes but using the sampling-time procedure adopted with phototransistors. The results of this third set of experiments were consistent with those of the two previous ones. By this statement we mean that the average value of the differences on detector A in the two lighting situations of the sources was below the threshold energy for the breakdown of LLI for the electromagnetic interaction, as required by the theory. In particular the maxima of $|\Delta \mathrm{A}|$ accumulate around the value of $2.3\,\mu\mathrm{V}$, in agreement with the results of the other two sets of experiments and again with the results from the laser cross beam experiment. Let us note that the photodiodes used as detectors in the first and third sets were integrated to a transimpedance amplifier, transducing the photocurrent signal into a voltage signal. Such a voltage, measured by means of a multimeter, does not depend therefore on the value of the circuit resistances of the voltage measuring system.

One can conclude that the sampling time procedure, which permitted the effect to be evidenced on phototransistors, could reveal it on photodiodes as well.

Therefore, it is possible to state that, although there is not such a tight bond between detector and time procedure, the latter plays a very important role in giving evidence to the effect. More explicitly, we could say that the phenomenon, when trying to detect it and to study, possesses so complex features, which make it hard to be grasped both in literal and figurative sense. We already know that this anomalous interference manifests itself only under precise conditions, namely below an energy threshold and within some spatial threshold as well. In this sense, it is endowed with a peculiar structure both in energy and in space. The global view of these three sets of experiments teaches us that there exists also some sort of threshold for the sampling time interval. Because of this time structure, the effect looks quite different depending on the time procedure adopted to sample the signals on the detectors, as it is apparent from the two responses got from the first and the third sets of experiments. In order to evidence this anomalous photon behaviour, which is the consequence of a very complex physical phenomenon, including the violation of the LLI, one has to adapt the physical inquiry to it and be aware of the existence of these thresholds in energy, space and time.

We want to add that the third set of experiments was repeated several times over a whole period of four months in order to collect a fairly large amount of samples and hence have a significant statistical reproducibility of the results. Thanks to this large quantity of data, it was possible to study the distribution of the differences of signals on detector A, which is shown in Fig. 14.7. For the sake of clarity, we reported only the differences ΔA outside the interval $[-1, 1]$, which is the interval of compatibility with zero of the values of ΔA.

The circumstance that the majority of the differences ΔA is negative, namely A(S1 on S2 off) $<$ A(S1 on S2 on) might wrongly induce to deem that, when source S2 was turned on, the signal detected on A increased.

One might then be incorrectly tempted to account for this by stating that some photons of S2 passed through aperture F2, see Fig. 14.6. Conversely, if one takes into account the mode of

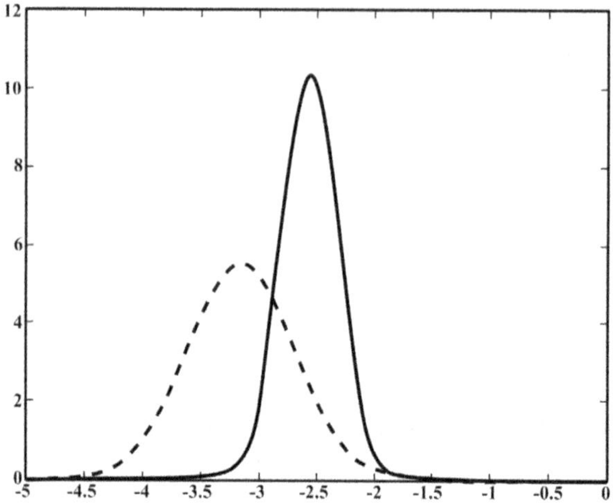

Fig. 14.7. Gaussian curves, normal frequency vs. signal difference in μV, for the signal differences ΔA and ΔA$'$ on detector A for the two cases of source S2 off and on, dashed and solid curve respectively. The instrumental drift has been taken into account. It is $\Delta A = -3.15$ ($\sigma = 0.45$); $\Delta A' = -2.56$ ($\sigma = 0.24$).

operation of the photodiodes chosen as detectors, it becomes immediately apparent that the above inequality means exactly the opposite situation. In order to understand this point, let us give some brief details about the operational mode of the type of photodiode, model OPT301 Burr Brown, used in the third set of experiments. First of all, we have to say that its pins were connected to the input pins of a trans impedance operational amplifier which was integrated along with the photodiode on the same chip. The photodiode was not inversely polarized and the dark current was always greater than the photocurrent. As is well known, the two currents flow in opposite directions, and the total current flowing in the photodiode is given by their subtraction. When the total current increases, the op-amp output voltage increases too. However, a rise of the total current, and hence a rise of the output voltage, means a decrease of the photocurrent, as the dark current cannot change, and this means a drop of the number of photons received by the photodiode. Thus, when both sources were on, the increase of the output voltage means that the photodiode A was receiving less photons. Namely, when S2 was turned on, detector A recorded a lower signal and hence received less photons, although there was a larger number of photons in the box because both sources were on! Needless to say, the stability of power supplies was constantly checked. On the other hand, it is impossible to account for this reduction of the signal on A when S2 got turned on and S1 was already on as a destructive interference between photons from the two sources, because the LEDs are incoherent sources of light.

In order to further enforce the evidence of the difference between the two physical situations it was carried out a statistical analysis of the results found in the two cases, again only the differences ΔA outside the interval $[-1, 1]$ have been considered, also taking into account the instrumental drift. The Gaussian curves obtained are shown in Fig. 14.7. The dashed curve refers to the signal differences ΔA = A(S1 on S2 off) − A(S1 on S2 on), whereas the solid one to ΔA′ = A(S1 on S2 on) − A(S1 on S2 on). The Gaussian curves, normal frequency vs. signal difference in μV, show the signal differences ΔA and ΔA′ on detector A for the two cases of source S2 off and on, dashed and solid curve respectively. The instrumental drift has been taken into account. It is $\Delta A = -3.15$ ($\sigma = 0.45$), $\Delta A' = -2.56$ ($\sigma = 0.24$). The two curves differ by 2.5 σ, clearly showing that the two cases are statistically distinct, the latter one representing a mere fluctuation unlike the former.

We can therefore conclude that the results obtained on the anomalous behaviour of photon systems brings to light a more complex physics of the electromagnetic interaction, which again calls for giving up the LLI in order to be accounted for. They are apparently at variance with both standard quantum mechanics, in the historical Copenhagen interpretation, and usual electrodynamics both classical and quantum.

The interpretation in terms of DST theory is quite straightforward. Under the energy threshold $E_{0,\text{e.m.}} = 4.5\,\mu$eV, the metric of the electromagnetic interaction is no longer Minkowskian, see Fig. 14.5. The corresponding space time is thus deformed. Such a space time deformation shows up as an 'hollow' wave accompanying the photon, and is able to affect the motion of other photons. This is the origin of the anomalous interference observed. We can note, according the DST theory, that the difference of signal measured by the detector A in all the double slit like experiments can be regarded as the energy spent to deformed space time. In space regions where the external electromagnetic field is present, the regions of 'standard' photon behaviour, we can associate such energy to this difference and what could be called 'hollow' wave is nothing but a deformation of space time geometry. By a metaphoric image we may picture the deformed space time, which is intimately bound to each photon, as the 'shadow' of the photon.

It is immaterial, like a shadow, since it carries neither energy nor momentum, and it can fill space regions far from the photon, exactly as a shadow fills space regions far from the body that casts it.

The space time deformation spreads beyond the border of space and time sizes, which correspond to the photon wavelength and period respectively. This changes the photon-photon cross section and therefore similar effects might be observed in photon-photon interactions, for instance in crossing photon beams. between the energy density corresponding to the external e.m. field and that of the deformed one. But also in electron-photon interactions as seen in the inverse Compton scattering.

14.4 Angular Scheme

Up to now in the two precedent paragraphs we dealt with the energy part of the LLI violation. In particular the cross beams experiments could not give anything else, despite the fact that the different pattern of the laser beam spot recorded by Charged Coupled Device (CCD) image detector could allow a match between spatial changes in the pattern of the spot images and the related measure of energy in the whole image divided in pixels. Now using the experiment of the 'shadow of light' we can perform an angular analysis of this phenomenon and make the comparison with the angular behaviour of the LLI violation as seen in Chapter 13. In Fig. 14.8, we resume the angular scheme valid for both boxes using as detectors photodiodes or phototransistors, even if in this picture, for practicality, reported is the layout according to mirrored Fig. 14.6. In this angular scheme we take into account the fact that the 'shadow of light' effect occurs only for detectors frame within 4 cm far from the circular slits plane. In the experiments the sets of measurements were taken for distances of 1 cm step and in Fig. 14.8 are shown the angular sketch for the last two positions numbered 3 and 4 corresponding to distances 3 cm and

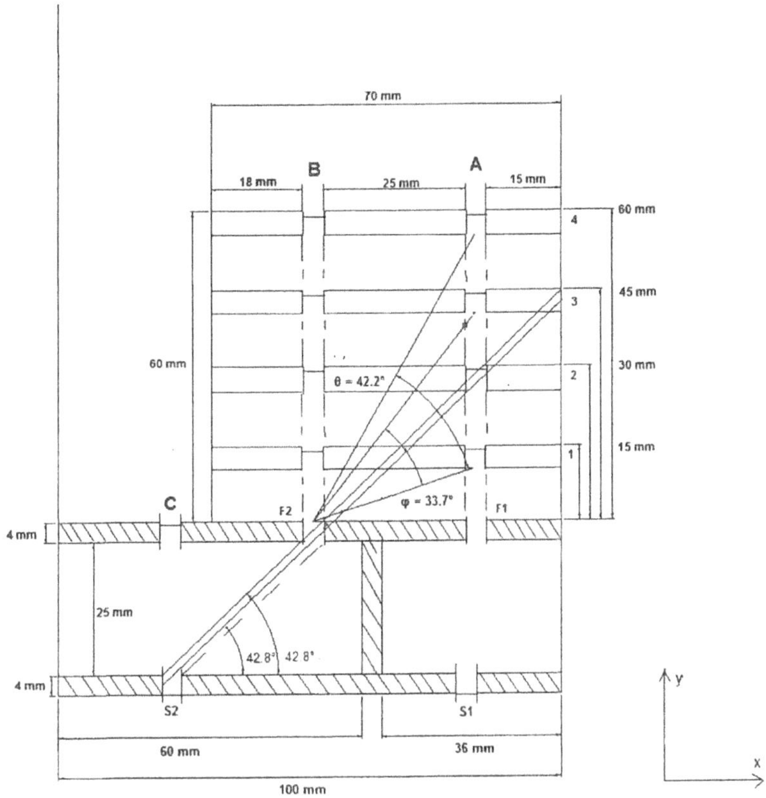

Detectors A,B,C Sources S1,S2 Slits F1,F2 Ø 4,5 mm.

Fig. 14.8. Lorentz violation angle for the shadow of light effect.

Notes:

$\varphi = 33.7°$ Opening angle for the come within which the photons coming from S2 crossing F2 affect the photons of S1 detected by detector A

$\varphi = 33.7°$ position 3 cone opening within the shadow of light phenomenon occurs

$\theta = 42.2°$ position 4 cone opening within the shadow of light phenomenon occurs

The LLI violation angles in the magnetic coil were $\varphi = 0.615\,\text{rad} = 35.26°$ and $\theta = 45°$ the angles are reported in sessadecimal units.

4 cm, for higher distances the phenomenon disappears. As well evident there is match between the angles of LLI violation, see Fig. 13.6 and Fig. 13.7, and the angles of the cones within which there is the

spatial region where the shadow of light produces its effect in photon-photon interaction see Fig. 14.8. In fact the angles coming from the shadow of light are within the corresponding opening values of the angles coming from magnetic coil experiment. Therefore, we can conclude that there is an 'angle law' underlying all the two phenomena examined up to now which is intimately connected with the LLI violation and space time deformation. This fact gives further support to the evidence that the LLI violation is due to geometrical asymmetry and its origin is geometrical not kinetic.

References

[1] F. Cardone. *Nuovo Cimento A*, 104A(5), 757, 1991.
[2] A. Ranfagni *et al. Phys. Rev. E*, 69, 027601, 2004.
[3] A. Ranfagni *et al. Phys. Rev. E*, 69, 027601, 2004.
[4] R. Meucci *et al. Modern Phys. Lett. B*, 19, 1, 2005.
[5] L. de Broglie and J. Andrade y Silva. *Phys. Rev.* 172, 1284, 1968.
[6] F. Cardone *et al. Phys. Lett. A*, 326, 1, 2004.
[7] R. Scrimaglio *et al. Int. Jour. Modern Phys. B*, 20, 85, 2006.
[8] W. Perconti *et al. Int. Jour. Modern Phys. B*, 20, 1107, 2006.

Chapter 15

Angle Law: Torsional Antenna

The previous two chapters, Chapters 13 and 14, examined the 'angle law' in a passive way, it means that the experiment performed 'found' this behaviour and tested it in two different physical systems, one using static field and the other using photons. Effectively they performed a cross-check test of this law which has been found. Now in order to perform a double cross-check test the angle law is explored in an active way, or rather starting from the knowledge of the angle law as main feature of the local Lorentz invariance (LLI) violation in order to investigate the behaviour of another physical system different from the two already used. This way has two main goals. First to verify once more this law performing what can be called the 'repeatability in great', which means to repeat the occurrence of a law ruling in the same way different phenomena in different physical systems, even ruled by the same interaction, giving rise to the interpretation that this law is fundamental part of the interaction or an independent 'natural law'. Second to test the behaviour of the energy of the physical system used in order to check by experimental way if the 'lost symmetry' due to LLI violation can be recovered producing a symmetry in the energy generated by physical system under the constraints of the angle law.

Following the hints attainable from the two systems used, magnetic coil and emitted photons, it was used a system emitting photons but under the 'deformation' of the emitting device. The chosen system was a 'torsional antenna' emitting electromagnetic waves and the test was the polar diagrams of the emitted energy at different torsion angles.

In order to investigate the effects of the torsion angle, a systematic study was performed [1,2], in particular, the signal of a transmission antenna as a function of the torque angle was investigated, the not torqued case corresponding to a so-called 'cross-loop antenna'. In the following, we present and discuss the results obtained when a log-periodic antenna is used to study the emission of this torqued antenna.

Examples of antennas based on torsion around an axis are the Helica Filar boards (QHF), suitable for receiving the transmissions of the NOAA-sat polar satellites and WX-sat on the frequency of 137 MHz [3]. These antennas were designed in the United States at laboratories of the National Air and Space Administration (NASA) and are used in any area where a good sensitivity and an appropriate directionality are pursued. They consist of square loops twisted in a spiralling shape to form an apparent torsion angle of 180°; the presence of a balun and their stiffness make them different from the one we are considering and not so simple to put in evidence the effects of twisting the two loops. Moreover, the influence of surface twists on the characteristics of an antenna, to be placed on the nacelle of military helicopters, was studied [4, 5].

Therefore, taking into account all this knowledge, the torsional-antenna used is a quadrupole, made of two concentric circular dipoles arranged on orthogonal planes. The variable torsion angle is measured with respect to this condition, in which the dipoles are on orthogonal planes. In the torqued case, we studied, the envelope is an ellipsoid of rotation. On the contrary, in the twisted case the envelope is a cylinder, or a cone, this case corresponding to helical or conical antennas which is not matter of the performed experiment and study. As said above in this experiment, the working hypothesis was that, if any anomalous behaviour was observed, it could be put in correlation with the behaviour observed in the case of electromagnetic interaction when a resistance/capacitance circuit inside a static magnetic field was studied at different directions giving rise, as seen, to the evidence of LLI violation according the angle law. In the case described in Chapter 13, an unexpected voltage was observed both in the horizontal plane and in the vertical plane perpendicular to the North direction of the terrestrial magnetic field. The magnetic North direction was chosen as conventional reference direction and later on used in all the experiments and measurements which tested the LLI violation. Therefore the conventional Cartesian orthogonal reference frame used was the same as in Chapter 13 where the wider Lorentz violation occurred in the vertical plane at $5\pi/4$ rad.

The investigated antenna is made of two vertical loops, mutually perpendicular, having almost circular shape. In fact, each loop is made of two semi-circular branches with 16 cm radius. Each semi-circular branch is kept at 15 mm apart from the central vertical axis by using two PVC cubes, one on the top and one at the bottom part. The two branches of each loop are made of harmonic steel, see Table 15.1, having circular section with 0.5 mm radius, and are closed on the top through a horizontal conducting connection of 30 mm which crosses the PVC upper cube. These two perpendicular connections are at a distance of 2 mm each other, thus avoiding the contact between the two loops. As a consequence, each loop is 100 cm long and 32 cm high. The bottom of each loop is connected to a channel of a signal generator, HP 8510 Vector Network Analyzer.

The torqued configurations were obtained by keeping fix the lower part and rotating the upper PVC cube. The torsion angle of the upper part will be indicated as the torque angle α.

The configuration corresponding to torque angle $\alpha = 180°$ is visible in Fig. 15.1.

The two loops of the antenna were separately power supplied by using the two output ports of an HP 8510 Vector Network Analyzer, with no phase shift, a power of 1 mW each on an impedance of 50Ω and a frequency of 240 MHz. Two sets of measurements were performed: one with antenna axis vertical, see Fig. 15.2, and one

Table 15.1. Antenna wire composition.

Carbon steel – EN10270-1 class SH, DIN17223, ASTMA228	
C	0.806%
Si	0.206%
Mn	0.432%
P	0.013%
S	0.004%
Cu	0.028%
E	206 kN/mm^2
G	81.5 kN/mm^2
ϕwire	1 mm

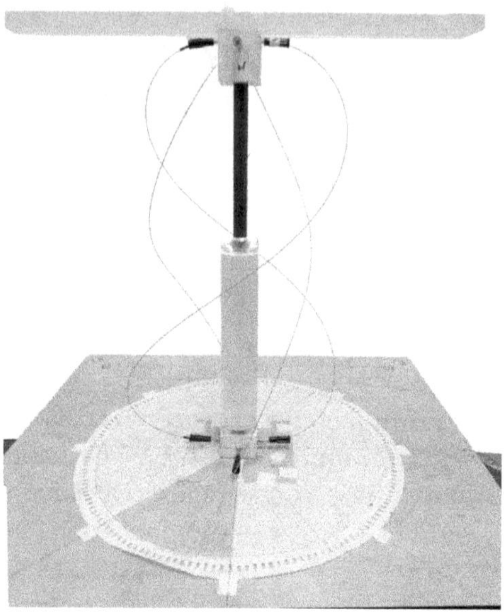

Fig. 15.1. The torsional antenna at a torque angle $\alpha = 180°$, vertical configuration.

with the axis horizontal, see Fig. 15.3. At a distance of 220 cm, a log-periodic antenna was placed as a receiver of the emitted signal [2]. This log-antenna was kept fixed, while the emitting torsional-antenna could rotate on a table around a vertical axis crossing the center of the antenna, rotation angle ω, thus keeping the α torque-angle fixed. The total range of the rotation angle ω was 360° clockwise, with 180 steps of 2°. At each ω step, intensity was measured during 1 s. In this way, the relative rotation of the log periodic antenna was anticlockwise. Particular attention was paid to the orientation of the apparatus with respect to the North of the Earth magnetic field: to this aim, a Cartesian coordinate system was chosen having the z axis vertical and the y axis in the direction of the magnetic North (±2°). With this choice, a loop was in the xz plane and one in yz plane when $\alpha = 0$ and $\omega = 0$ and axis of the antenna was vertical in the vertical configuration, see Fig. 15.2, with the transmitting antenna being in the vertical configuration the units of the received signal, expressed in dB, are reported as absolute value, in the range [−60, −20] dB. Eight values of torque angle α are considered: 0 rad (a), $\pi/4$ rad (b),

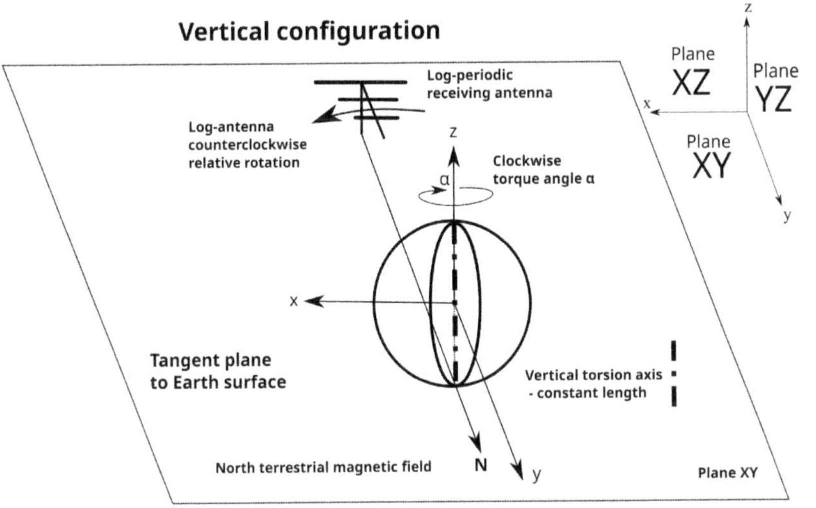

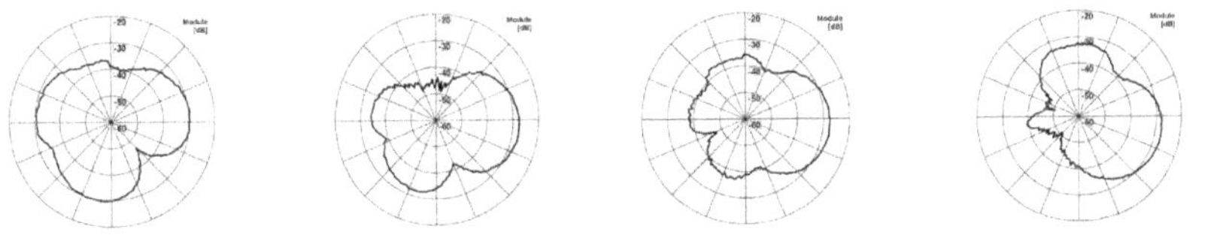

(a) 0 rad Torque = 0° h: 9:55 – 9:58

(b) π/4 rad Torque = 45° h: 9:59 – 10:02

(c) π/2 rad Torque = 90° h: 10:03 – 10:06

(d) 3π/4 rad Torque = 135° h: 10:07 – 10:10

(e) π rad Torque = 180° h: 10:11 – 10:14

(f) 6π/5 rad Torque = 215° h: 10:15 – 10:18

(g) 5π/4 rad Torque = 225° h: 10:19 – 10:22

(h) 3π/2 rad Torque = 270° h: 10:23 – 10:28

Fig. 15.2. Vertical configuration of the torsional-antenna: antenna torsion axis perpendicular to horizontal plane. Polar representation of intensity registered by log-periodic antenna as a function of torque angle α.

$\pi/2$ rad (c), $3\pi/4$ rad (d), π rad (e), $6\pi/5$ rad (f), $5\pi/4$ rad (g) and $3\pi/2$ rad (h).

In the horizontal position, antenna axis was along the y direction. The $\alpha = 0$ and $\omega = 0$ configuration correspond to loops in the xy and zy planes, see Fig. 15.3 showing the horizontal configuration, with the transmitting antenna being in the horizontal configuration,

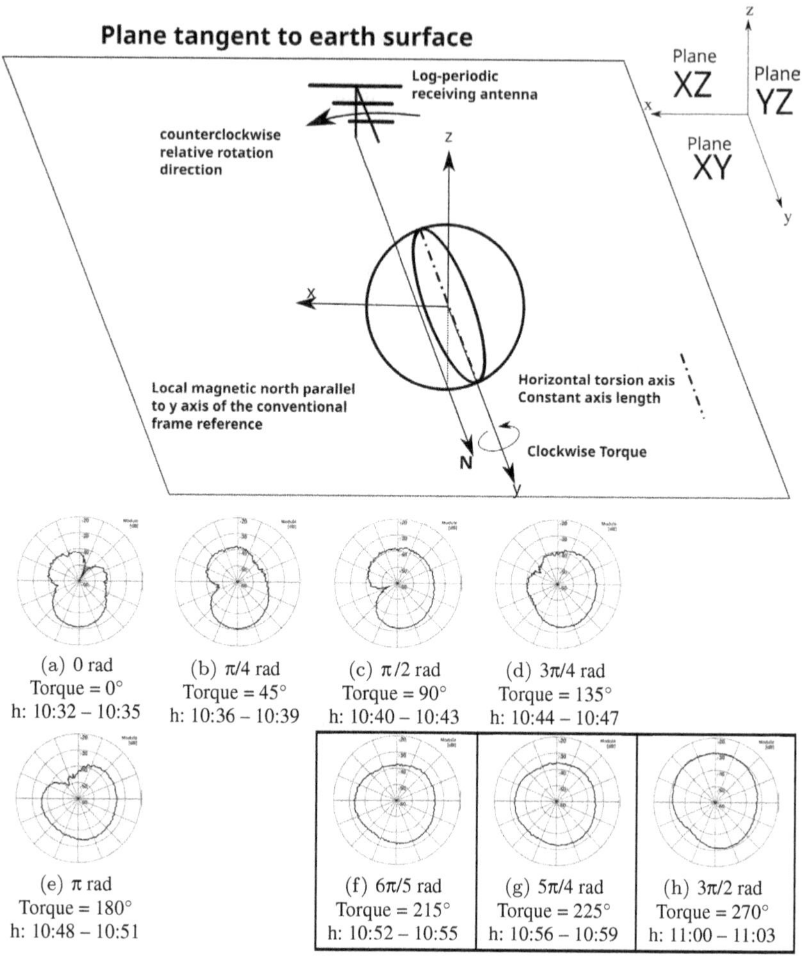

Fig. 15.3. Horizontal configuration of the torsional-antenna: antenna torsion axis parallel to horizontal plane and laying along the North direction. Polar representation of intensity registered by log-periodic antenna as a function of torque angle α.

the units of the received signal, expressed in dB, are reported as absolute value, in the range $[-60, -20]$ dB. Eight values of torque angle α are considered: 0 rad (a), $\pi/4$ rad (b), $\pi/2$ rad (c), $3\pi/4$ rad (d), π rad (e), $6\pi/5$ rad (f), $5\pi/4$ rad (g) and $3\pi/2$ rad (h).

Figure 15.2 reports the polar diagrams of the intensity registered as a function of angle ω when the antenna was in vertical configuration, Fig. 15.3 refers to horizontal configuration.

The most evident effect of torquing antenna is that the intensity distribution is almost isotropic at torquing angles α larger than π rad in the horizontal configuration, see Fig. 15.3, polar diagrams f, g and h.

Before performing the experiment, one should have expected the most isotropic distribution at large torquing angles in the vertical configuration, not in the horizontal one. In fact, in the former case, if one considers the antenna as an assemble of many dipoles, their directional distribution is more homogeneous at high torquing angles, thus creating an isotropic signal in a horizontal plane, which is the plane of the receiving antenna reported in Figs. 15.2 and 15.3.

In order to check this point, a three dimensional model of the antenna was created by using a numerical model, which has been implemented in MATLAB software. The final simulation of the electro-magnetic signal, as expected by applying standard software based on well-known classical Maxwell equations, was performed by ANSYS HFSS software.

No such effect of isotropy at high torquing angles in the horizontal configuration was obtained and anyway no isotropic distribution has been obtained at all. From these simple considerations one can deduce that the registered isotropy is not a direct consequence of the geometry, at least of the geometry we are acquainted to in a flat Minkowskian spacetime.

As said the explanation for this behaviour was looked for by considering the reported behaviour of a static conductor inside a static magnetic field according to the 'angle law'.

We recall here below the main features of the magnetic coil experiment where it was used a conventional Cartesian coordinate system with y axis along the terrestrial magnetic North and z perpendicular to the local earth surface and the following effects were observed.

In the yz-plane, inside the experimental error, no voltage was detected at different directions of the conductor, as it is expected.

In the xy-plane, inside the experimental error, no voltage was detected at different directions of the conductor, as it is expected, apart from an unexpected voltage at $\pi/4$ rad, but not at 0 rad nor at π rad, from the North. It was observed during the measurements from 9 to 9:59 am of the 8th of June 1998 but not between 0:00 and 0:46 pm of the same day.

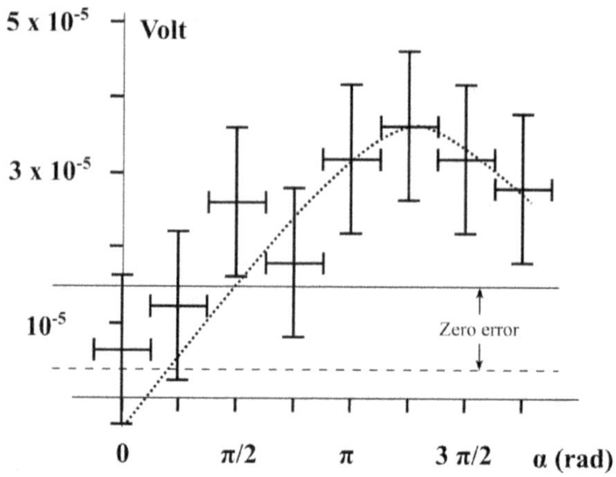

Fig. 15.4. Voltage detected in a static conductor inside a static magnetic field as a function of α angle formed with respect to the vertical, in a plane perpendicular to the North direction of the local magnetic field.

In the xz-plane, a wide variation of the voltage as a function of α angle from the vertical direction was detected, as depicted in Fig. 15.4.

$$\alpha_{\max} = 3.757\text{rad} \cong 6\pi/5\,\text{rad}$$

The instrumental zero corresponded to $(1.5\pm1.0)\ 10^{-5}$ Volt. Thus, only the measurement at $\alpha = 5\pi/4$ is out of zero range, the error in each measurement being $\pm1.0\ 10^{-5}$ Volt.

However, as the zero error is systematic, whatever is its assumed value, many results are different from zero. Moreover, the results are not randomly distributed, but correspond to a continuous variation with angle, as shown by the dotted line in Fig. 15.4. The central value of the so obtained peak is about $6\pi/5$, after applying correlation tests and interpolating procedure.

In particular, if the central value 1.5×10^{-5} V is assumed as zero value, a wide region, at α larger than π rad, corresponds to the occurrence of a voltage which must not exist, ever according to known behaviour of the electromagnetic interaction without LLI violation.

The results reported in Fig. 15.4 were obtained between 8:30 am and 9:30 am of the 9th of June 1998. A following measurement between 12:18 pm and 13:05 pm of the same day did not confirm this occurrence of an anomalous voltage. These results have been confirmed in the next test runs performed in the years 1999–2000.

Both the results in the xz-plane and in the xy-plane show a dependence from the time of the experiment. Due to this fact, a dependence from a direction different than the North magnetic field was suggested. In particular, some cosmological direction, which could be related to some anomaly of the microwave background radiation and its cold spot preferred direction was proposed, as we will see in the Part V of the book.

At any rate, these results correspond to a breakdown of the local Lorentz invariance (LLI), as different results are obtained in different directions and in different times.

This anomaly occurs in directions corresponding to those torsion angles of the antenna for which the detected signal is isotropic, or rather angles greater than π with respect to the vertical in a plane perpendicular to the North direction. In this angular range, a violation of the LLI occurred in the case of the conductor inside the static magnetic field and, according the deformed space time theory (DST) this violation is attributed to metric element, energy dependent, following the energy thresholds of each interaction.

Thus, outside each threshold, the space-time is not flat for all interactions at all energy values, the non-isotropic and non-symmetric space-time corresponds to a violation of LLI and, as a consequence, to a breakdown of the same Maxwell equations for electromagnetism. Therefore, in the case of the conductor, a deformation of space-time was detected which can be put in direct correlation with the case of the antenna, given that a deformation exists, or rather the torsion of the antenna counteracts this deformation and makes the energy signal isotropic.

The followed procedure to recover the symmetry is a step ahead with respect to the already followed methods. In fact, when the three dimensional (3D) translation symmetry of the Galileo relativity was broken, by the existence of the Lorentz force, a new symmetry was found by adding the time as a fourth dimension and the four-dimensional (4D) Minkowski space-time was introduced. From the 3D symmetry of the Euclidean-Cartesian space, x, y and z spatial directions, whereas t is the Newtonian absolute time, a 4D symmetry of the pseudo-Euclidean Minkowski space-time (x, y, z, t) was obtained. In the present case, the violation of 4D Lorentz symmetry was recovered by adding a new variable, the energy, which in turn is related to the angle of Lorentz violation, thus, a 5D space-time-energy (x, y, z, t, E) is needed and then obtained. From the methodological point of view, the two steps are different. In fact, the transition from 3D to 4D is a mathematical way rearranging some equations to obtain in 4D the symmetry which is lost in 3D. On the contrary, now the isotropy was obtained experimentally, thus underlying a transition from 4D to 5D which was studied in detail from the mathematical point of view using the method of the Ricci flow as discussed in Part II of the book.

We resume this procedure in Fig. 15.5, where we observe that the polar distribution of the emitted, or irradiated, energy assumes a near-isotropic and near-symmetric shape and profile at higher deformation of the emitter according and in perfect analogy with the corresponding geometry of the Lorentz violation as detected in the magnetic coil experiment. The plane is the same as the vertical one (x z), the angle is the same as between the torsion of the antenna and the orientation of the detected violation. The conclusion is that according to the angle law, stating the asymmetry of Lorentz violation giving a deformed space time in five dimensions, when we deform the device emitting energy we obtain the recovery of the asymmetry, due to the violation, in the symmetry of the fifth coordinate, the energy.

We noticed that the use of the energy as fifth physical coordinate allows to recover the symmetry and to solve the Lorentz violation, but the use of the fifth coordinate energy sets three problems we must face and deal with.

First, we have to deal with the energy in a separate way for each interaction, taking well in account that we measure the energy only by electromagnetic interaction. Therefore, this first problem can be summarized by the question on how the energy measures the energy.

Second, which form assumes the fifth metric parameter corresponding to the coordinate energy, again this problem is evidently connected to the question on how the energy measures energy.

Third, the energy is a dynamical variable: we must examine the eventuality of extending the role of dynamical variable also to the other four coordinates, the three spatial ones and the time.

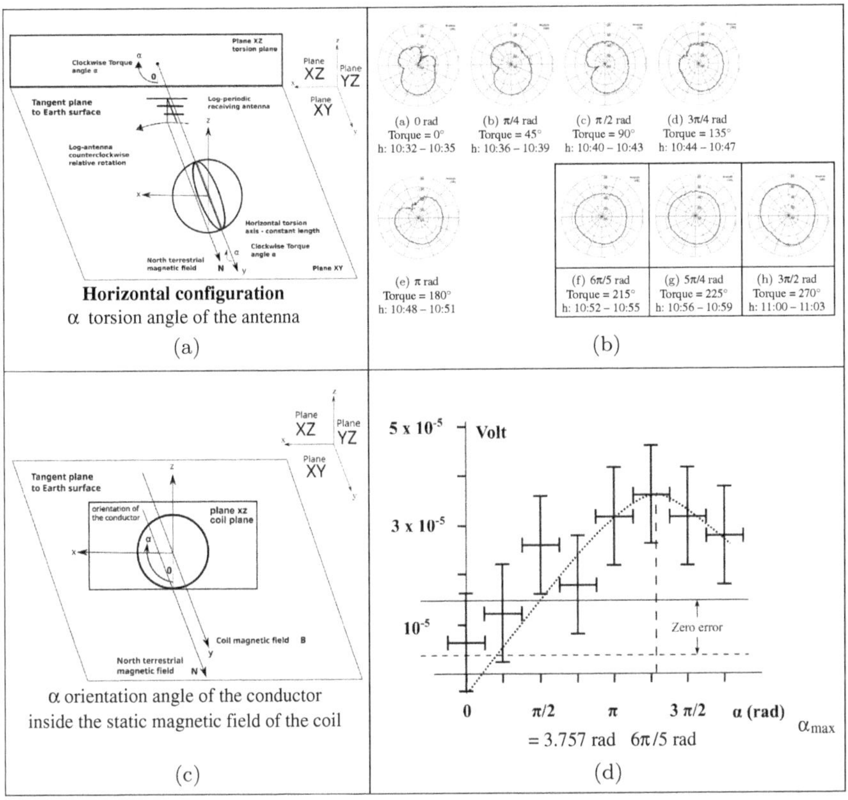

Fig. 15.5. (a) Sketch of the torsional antenna in horizontal configuration. (b) Eight polar diagrams of radiation on meridian plane with respect to the horizontal torque axe. (c) Sketch of the magnetic coil test for LLI violation. (d) Voltage detected in a static conductor inside a static magnetic field as a function of α angle formed with respect to the vertical, in a plane perpendicular to the magnetic North direction.

This third problem implies that a geometrical representation of the interactions by metric parameters depending on the energy, intended as phenomenological variable, pave the way to consider all the coordinates as dynamical variables. At last, we remark that the energy has a double role in the framework of such a five-dimension metric: it is a coordinate and a phenomenological variable, on which the metric coefficients depend. In the framework of this double role, the question on 'how the energy measures the energy' finds a solution in the metric element of the coordinate energy. Therefore, it is possible to put forward the 'ansatz' that this fifth coefficient of the metric for each interaction plays the role of the gauge for the energy. All these three issues found the solution in the fifth element of the energy metrics of the interactions as illustrated in the Part II of the book. Now we do a backward step and show how the polar diagram of the neutrons emitted by deformed space time (DST) reactions, as described in Chapter 6, assumes the shape, as shown in Fig. 15.6, which resembles to the asymmetrical polar diagrams of the emission of the torsional antenna, a sort of 'cardioid'. The issue is if it can be possible to manage this asymmetry in the hadronic interaction using

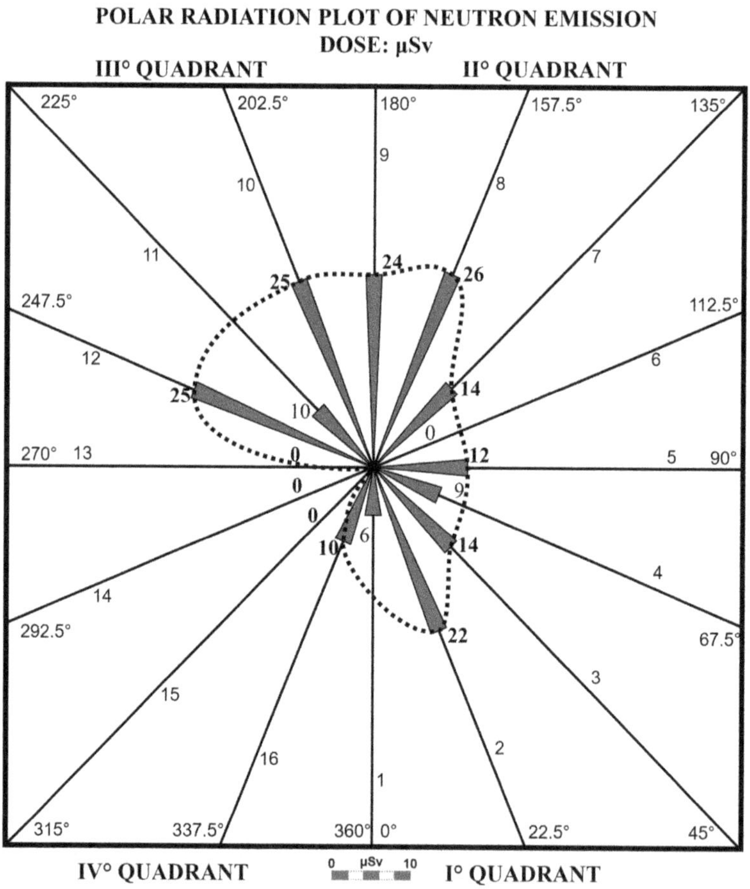

Fig. 15.6. Polar diagram of the neutrons emitted by DST reactions.

the angle law as done for the electromagnetic interaction, in order to obtain stable nuclear DST reaction at macroscopic level. This problem has a solution mainly on technological ground but based ever on the suitable use of the angle law.

References

[1] Patent n. ITBO020120598(A1)-2014-05-06.

[2] D. Paci. *Antenne di Torsione e possibili conseguenze nella Fisica Fondamentale* (in Italian), Ph.D. Thesis - Università Politecnica delle Marche – XVII Edition New Series, Ancona, 2018.

[3] C. C. Kilgus. *IEEE Trans. Antennas Propag.*, 23(3), 392–397, 1975.

[4] I. M. Kilgore *et al. IEEE Antennas Wireless Propag. Lett.*, 15, 1242–1244, 2016.

[5] G. S. Karthikeya *et al. Antennas and Propagation (APCAP), IEEE* 5th South Pacific Conference pp. 285–286. IEEE, 2016.

Chapter 16

Symmetric Energy

In general and usually, symmetry breaking and space time properties are kept separate. In fact when it is realized the need of symmetry breaking, this is handled as a spontaneous phenomenon independent from both space time properties and their local Lorentz invariance (LLI), which warrants and guarantees these properties. Essentially the usual space time properties connected with Lorentz invariance assert that the space is isotropic and the space and time are homogeneous. At least in the framework of the Minkowskian spacetime it is well known that these properties guarantee, via the Noether theorem, the conservation laws, say the space isotropy gives rise to angular momentum conservation and the space homogeneity accounts for the linear momentum conservation, in the meanwhile the time homogeneity accounts for the energy conservation, in Minkowskian language the time momentum conservation. Again it is well known that the independence of any symmetry breaking from both the space and time properties, and thus Lorentz invariance, produces the need to 'insert by hand' the symmetry breaking in the interactions description. The symmetry breaking is therefore 'justified a posteriori' from the experimental check of the physical consequences of such a breaking which, we stress once more, does not affect the Lorentz invariance. All these facts bring us to the following consequences. We face the physics of phenomena which shows both space time anisotropy and a new form of Local Lorentz Invariance violation ruled by the angle law, therefore we can speak of a 'directional Lorentz violation'. Even if we manage this situation by using the energy in the double role of phenomenological variable of the metrics of the interactions, the energy metrics, and as fifth dimension in the energy metrics, which is the core of the deformed space time theory, the problem remains, at least, of how handling the concept of homogeneity between space and time, that we already used in Chapter 10. We split this concept in the following two parts, the 'principle of homogeneity' of the space time and the 'property of homogeneity' of the spacetime in its usual meaning. The principle of homogeneity fixes the need to measure all the coordinates, used in the metric of an interaction, by the same physical unit. We used this principle choosing the unit of energy so obtaining the recovery of the symmetry through the symmetric energy distribution as we did in Chapter 10. This principle of homogeneity is well suitable and inserted within the framework of the theory of the deformed space time. On the other hand the property of homogeneity of the spacetime in the usual interpretation requests that in general all coordinates 'must' be measured in space unit and it is connected in the standard way with the two momenta conservation where the most important is the time momentum, the energy conservation. We prefer considering the energy conservation according to the Hamilton theorem, where the Hamiltonian function represents the total energy, instead of the use of Noether theorem which uses the Lagrangian function which is a functional of the energy and does not represent anyway the total energy. In this sense even in an anisotropic and asymmetric space time the total energy conservation is apparently violated but in fact preserved by the energy absorbed in the space time deformation. So the space time is no longer a static stage like the Minkowskian spacetime where the phenomena occur leaving it intact, or like the Riemannian spacetime where its geometrical properties influence the phenomena, like the curvature of light trajectories or the time delay as verified by the Shapiro experiment [1]. In these two cases the spacetime is in a certain sense separate and distinct from phenomena. Instead, in deformed space time, the space and time participate together with the interaction to the phenomena since they 'are' part of the phenomena because the space time energy metric 'is' the interaction. In this way, and not by hand, we consider both the symmetry loosing and the symmetry recovering. We could make a parallelism, maybe inappropriate, between these two couple of concepts: the spontaneous breaking of symmetry is substituted by the loss of the symmetry due to a fundamental asymmetry underlying the physical phenomena, the angle law, and the Lorentz invariance breakdown is substituted by the geometric directional Lorentz violation.

16.1 Ellipsoids Reduced to Sphere

In Chapters 9 and 10 we discussed the representation of the energy 'volume' together with the representation of the so called Cauchy tetragon inscripted inside this volume. The sense of this representation also thru the Cauchy tetragon is to show that the relevant angles of the tetragon may be connected with the 'angular violation' of LLI, and this tetragon can be put in correlation with the 'stress' giving rise to the space time deformation by the energy. In this sense we see the energy deforming space and time but it is as well absorbed, in part, by the deformation itself preserving in this way the total energy conservation, as already stated above. Here we do not wish to put forward further this ideas but only to explore how energy and symmetry are connected with the features of the 'geometrical Lorentz violation'.

We recall here the results of Chapter 10 or rather the representation of the energy 'volume' in the reference frame (z t E). We remind that axis z is along the direction of matter antimatter collision. As noted in the previous chapter we will see that the representation shall be symmetrical. In Table 16.1 we report the values of the semiaxes in GeV and in Fig. 16.1 is sketched the result, which is a spheroid since the semiaxes are nearly equal and therefore the Cauchy tetragon is nothing else than a cuboid.

It is intriguing enough that a suitable choice of coordinate axes together with the use of the homogeneity principle, and the choice of energy as the univocal unit of measure give rise to the possibility of a geometrical representation of the energy having the property

Table 16.1. Semihaxes of the spheroidal energy volume as sketched in Fig. 16.1.

Coordinate axis	Semiaxes	GeV
z	a_3	0.99
t	a_4	0.98
E	a_5	0.93

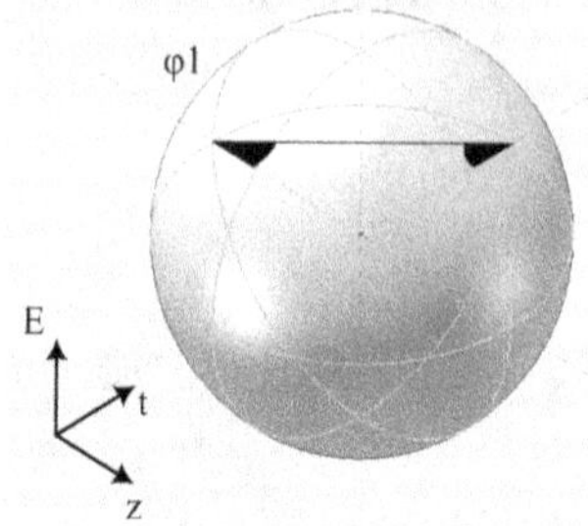

Fig. 16.1. Sphere of the annihilation energy the 'fireball' in the z, t, E space, the relevant angle connected with LLI violation is given in radians $\varphi_1 = \pi/4$.

to be symmetrical, or near-symmetrical, as a spheroid with low eccentricity. Now let us make some more considerations about this possibility to recover a lack of symmetry by using the energy symmetrization. We noted that representing the 'volume' of energy we moved from a geometrical entity, the non rotation ellipsoid, having high eccentricity, see Fig. 9.1, thru a rotation ellipsoid with lower eccentricity, see Fig. 9.2, then to a spheroid with the lowest eccentricity, see Fig. 16.1. We have to take into account that we followed this way starting from the hadronic metric having the property of anisotropy independently from the energy and therefore showing a violation of the LLI. The anisotropy of hadronic metric brings to a lack of symmetry, consequence as well of the lack of LLI. Now it is possible to state that a violation of LLI can be recovered introducing a new physical extended coordinate and performing a representation of physical phenomena by a metric space endowed with one more coordinate. As it was historically done when the Galilean symmetry was lost in electromagnetic phenomena, it was useful, or better necessary, to introduce one more coordinate, the time, and to change the Euclidean three dimensional geometry, which supplies the Galilean symmetry of the translations, with the Minkowskian four dimensional geometry, which supplies the LLI or rather the Lorentzian symmetry of the roto – translations. In our cases under consideration the violation of LLI represented by anisotropic metrics, if even asymmetric, is recovered by introducing a new coordinate, the energy, with a five dimensional geometry. We can state in general that when an invariance, like LLI, is violated, the first step to recover it is to introduce a new physical extended coordinate, not a simple mathematical one or ones. The problem is then its choice but the interactions give the clue. In order to recover the symmetry we are following the master way to get the symmetrization of energy distribution in space and moreover with respect to the angular distribution. As we will see in the following, this way can be better followed by experiments than mathematics. About the issue of the angular distribution of energy in the phenomena, we did the choice to borrow from the theory of elasticity the concept of Cauchy stress tetragon in order to evaluate the notable angles connected with hadronic metric which we will use in the next. More in general we saw, from phenomenological evidences, that the angle assumes a more and more relevant role through the angle law in describing both the lack of symmetry and the violation of LLI as far as being connected with LLI itself.

16.2 Cardioids Reduced to Circle

Here we recall the relevant results of Chapter 15 about the energy distribution we got from a torsional antenna. In Fig. 16.2 we report the polar representation of intensity registered as a function of torque angle α of the transmitting antenna for values of the torquing angle

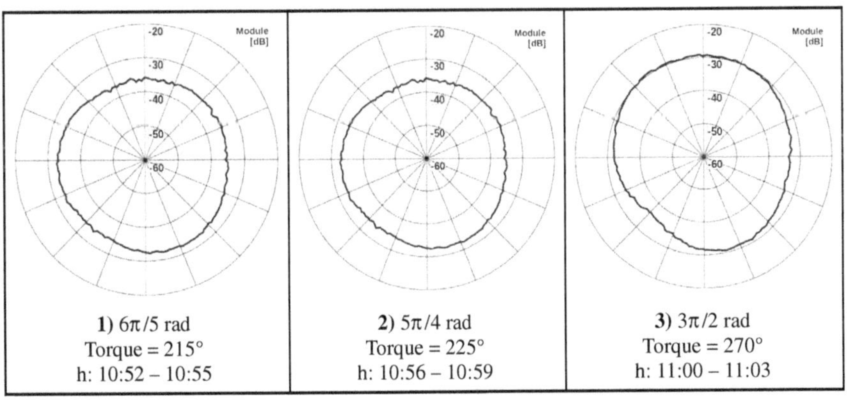

Fig. 16.2. Polar representation of intensity registered as a function of torque angle α of the transmitting antenna, the units are absolute value in the range $[-60, -20]$ dB. The three values of torque angle α considered are those giving a symmetric circular distribution of the radiated energy (1) $6\pi/5$ rad; (2) $5\pi/4$ rad; (3) $3\pi/2$ rad.

corresponding to interval of values matching the angle law of the geometrical angular violation of the LLI.

The results are symmetrical, or near-symmetrical, energy distribution having the shape of a circle with low eccentricity. In this case we do not have the necessity of any recourse to homogeneity principle.

Simply the situation of the phenomenon gives us the solution of the symmetry problem: using the angle law the energy distribution becomes symmetrical.

16.3 Sphere Circle Comparison

When comparing the sphere and the circle we must remind the different physical situation of the phenomenology, annihilation for sphere and radiation for circles, together with the deep difference of the energy ranges of the two phenomena.

But the most important difference is that for the sphere we resorted to the homogeneity principle, getting the clue of one of the LLI violation angle in the Cauchy tetragon thus connected to the angle law. Whereas for the circle we used physically the angle law.

Really they are the two faces of the same token, the procedure to recover the lost symmetry after LLI geometric directional violation passes through both the angle law and the homogeneity interpreted as the use of energy as common physical unit of measure for all coordinates, or rather the homogeneity principle. We will see how the geometric directional violation of LLI is intimately connected with an asymmetry which is independent from interactions and therefore underlying all the phenomena, as we shall see both at terrestrial level and at cosmological level.

Reference

[1] I. I. Shapiro *et al.* *Phys Rev. Lett.*, 20(22), 1265–1269, 1968.

Chapter 17

Lorentz Violation Recovered

The experiment of a static conductor, inside a static magnetic field, pointed out the evidence that violation of the LLI invariance occurred in some directions, defined by using the convention of a plane perpendicular to the north direction of the local magnetic field of the Earth. In the experiment of the torsional antenna the torsion angle α of the antenna exactly follows the same convention of the orientation angle α of the conductor inside the static magnetic field in the coil test for the Lorentz violation, as we have seen in all the treatments of the angle law. The symmetric distribution of energy in the polar diagrams exactly follows the Lorentz violation as it matches the same range of angle α. As a matter of fact, in the torsional antenna experiment the lost isotropy and symmetry are recovered by using as source of electromagnetic field a torsional-antenna torqued according the angle law. Once more we recall that the use of a Cartesian orthogonal reference frame with y-axis along the North direction of the Earth magnetic field is a pure convention without any physical meaning. We stress again that this conventional reference frame has been used in all experiments testing LLI violation, either for electromagnetic phenomena or nuclear ones.

Recovering the lost symmetry by using a torsional-antenna is a clear indication that the breaking, or better the violation, of LLI is mainly of angular nature. Therefore, the method of introducing a new physical dimension is a general rule to recover the lost symmetry and it is used here by introducing the energy as fifth dimension. We stress again that in the present case the recovery is studied experimentally, while it was introduced theoretically in the past by the use of new generalization of the mathematical structure, the Lorentz transforms, but on kinematic ground not geometrical one.

Now, the new issue is if the angle law, after its tests within the electromagnetic interaction, holds also for the nuclear interactions. Ever using the method of an experimental check we have to compare the goniometric reference frame, see Fig. 17.1, with the polar distribution of DST neutron emission, see Fig. 17.2, and thus test if they match together.

The problem is to quest if the angle law has a general value and is a natural asymmetry underlying all the phenomena at any scale. In this sense, we remark again that the energy has a double role in the framework of the five-dimension metrics, since it is a coordinate and a phenomenological variable, on which the metric elements themselves depend. In this framework of this double role, the question on "how the energy measures the energy" finds only a partial solution in the metric coefficient of the coordinate energy because we need also to

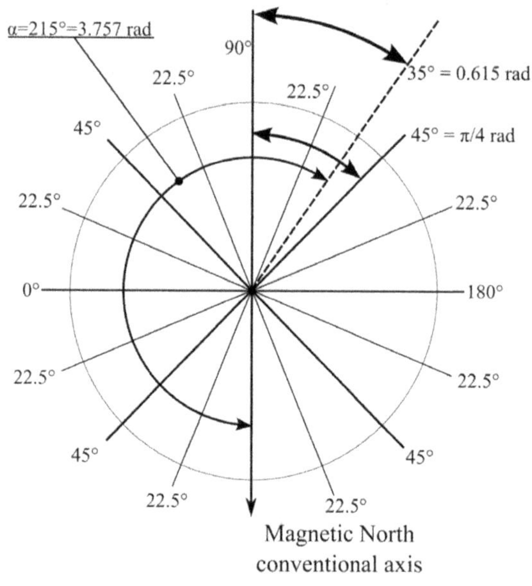

Fig. 17.1. Goniometric reference frame sketching the angles of the local Lorentz invariance violation, according to the experiment of the magnetic coil, α and $\pi/4$. Are shown the angles: $\alpha = 215°$ and its reduction to first quadrant or rather $\alpha - 180° = 35°$.

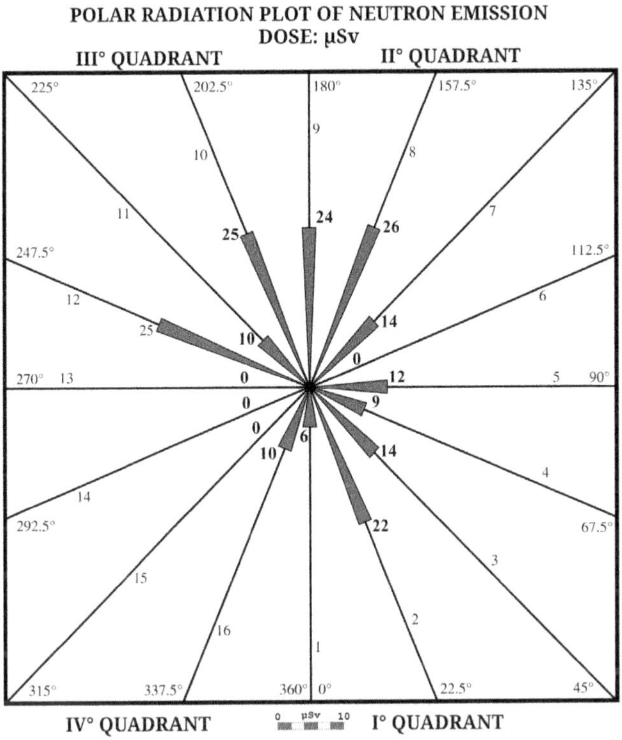

Fig. 17.2. Spatial Asymmetry of the neutron emissions from DST reactions.

take care of the angle law asymmetry in the mathematical representation of this metric element and possibly of all metric elements in order to include the spatial and temporal asymmetry.

Therefore, the ansatz, we already put forward, that this fifth element of the metric for each interaction plays the role of gauge for the energy, must be integrated with the request of its asymmetry, which means the loss of an universal gauge in the different directions of space as well for different astronomical times and unfortunately for different energy ranges too.

Now, we can conclude that the recovery of the Lorentz violation cannot elude the angle law from which itself comes.

Chapter 18

Nuclear Metamorphosis

18.1 Threshold Energy for Nuclear Metamorphosis

We briefly recall here the fundamental factors needed to build up any reactor producing in stable way nuclear metamorphosis. The reactor has to engage in the deformed space time (DST) reaction process a number of atoms, and their nuclei, comparable with the Avogadro number. To attain this result it must manage the so called energy densification at macroscopic level by using the indications from Eqs. (18.1)–(18.3).

$$D_{\text{had}} = 1.12 \cdot 10^{8}\,\text{J/m}^{3} = 112\,\text{Mega Pascal} \tag{18.1}$$

$$D_{\text{lep}} = 2.5 \cdot 10^{7}\,\text{J/m}^{3} = 25\,\text{Mega Pascal} \tag{18.2}$$

$$W = 7.6 \cdot 10^{11}\,\text{J/s} = 760\,\text{Gigawatt}. \tag{18.3}$$

Then the reactor must avoid the whole reaction becoming divergent and uncontrolled, but this can be obtained remaining in the 'flat region' between the two thresholds $E_{0,\text{lep}}, E_{0,\text{had}}$ of the metrics couple, which unfortunately is a very narrow condition but yet practicable. The last issue is the use of the 'angle law' within the macroscopic process of metamorphosis. The reactor using ultrasound must transfer the pressure through a sonotrode device whose shape has to be ruled by the results of the angle law and therefore avoid any 'lateral loss' of power through the lateral surface in order to get the power transfer concentration through its tip. Then the reacting materials receiving both power and pressure, if in liquid state, have to be 'shaped' according the angle law by using a suitable vessel, even if this implies an apparent violation of the D'Alembert paradox. The shapes of the device supplying the pressure and power and of the vessel where the material collects them are critical according to the angle law, because any change, even very small in value, is dramatically amplified by the power law of the fifth metric element of the hadronic metric which follows the sixth power of the energy. Therefore any change, say of factor ten, can amplify or depress the metamorphosis of a factor one million. When amplifying there is a risk to loose control of the whole reaction, but when depressing we get no metamorphosis at all. So, the building and the management of any 'reactor at metamorphosis' (RaM) must follow all these stringent prescriptions, which can be recapped as follows: there is only a narrow gauge on which the nuclear metamorphosis of matter runs and works. Last but not least we remind that the metamorphosis besides changing the matter is useful, at least at microscopical level, for producing bursts of DST neutron, as well as bursts of DST alpha.

18.2 The Nuclear Metamorphosis Reactor

The metamorphosis reactor, or RaM, can be described as a multi-purposes device producing the nuclear metamorphosis of the matter, as well as generating thermal energy and electrical charges during the reaction. The metamorphosis reactor was conceived and designed as a modular device intended to serve both as a test bed for different materials and their product as well as an industrial prototype testing several schemas [1]. The reactor provides thermal energy, coming from metamorphosis, with a coefficient of performance (COP) higher than 10. This COP comes from the thermal energy due to both electric charges and thermalized neutrons generated by matter metamorphosis produced by DST reactions.

An example of the thermic module applied to the reactor is sketched in Fig. 18.1. The modules for collecting electric charges and their electrical energy are made according to the energy harvesting methods which use the bucket brigade scheme, an example applied to the reactor core is sketched in Fig. 18.2.

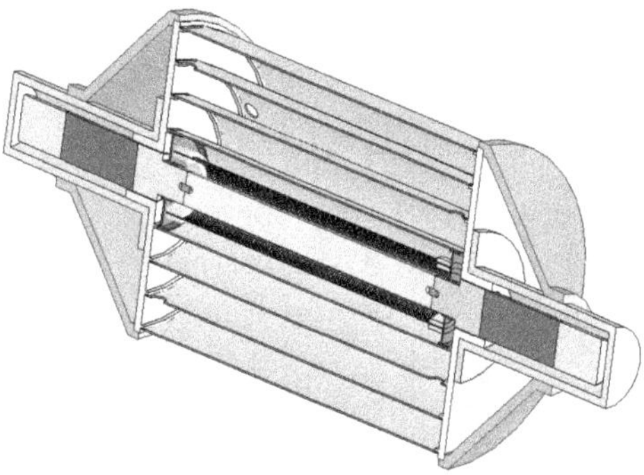

Fig. 18.1. Example of the thermic module for the metamorphosis reactor.

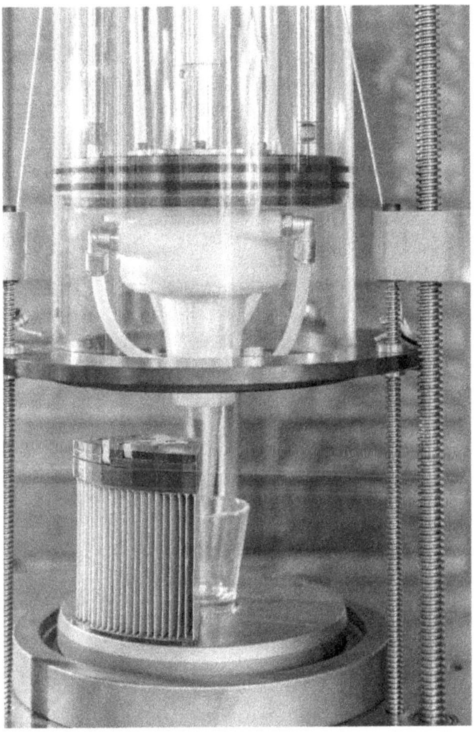

Fig. 18.2. An example of radial charges collector applied to the reactor core.

Now, we describe the modular structure and one of the uses of the metamorphosis reactor. The load-bearing structure is made by appropriately ribbed aluminium profiles to provide the system with high stiffness and stability. This structure ensures that spurious frequencies do not interfere with the correct production of the DST reactions in the reactor core. An industrial 2 kW power ultrasonic generator with a microprocessor and remote electronic controls was used for the reactor. The control system is composed of an electrical panel equipped with a microprocessor to drive the metamorphosis reaction during all its phases. Moreover the reactor is endowed with detectors of temperature, inner pressure, inner and external magnetic fields, as well as of inner and external electrical fields. All the data from these detectors are recorded in a central memory unit. This fully programmable reactor allows the user to repeat the same routine in the same way whenever wanted, and all the information on working conditions and sensors records are stored in the archive of the central unit. There is also a ground positioning system (GPS) receiver providing the astronomical as well as geographical position in order to obtain the relevant directions according to the 'angle law' for the DST reactions. The ultrasonic converter power unit used, equipped with an open-circuit cooling chamber, includes a booster and a sonotrode whose geometrical shape is designed according to the angle law. It is possible to change the whole ultrasonic unit to operate with either 20 kHz or 35 kHz. When applying the metamorphosis to a solid material it is also possible to use two coaxial ultrasound units at the two opposite ends of the sample. The ultrasonic unit including the sonotrode is thermo-controlled through an air cooling system, which allows to keep the working frequency indefinitely and therefore to obtain the 'energy densification' inside the sample undergoing the metamorphosis for a time interval at will. The reactor core is put inside a sealed and movable atmospheric chamber endowed with sensors of temperature and pressure. Moreover there are computer controlled inlet and exhaust solenoid valves allowing pre and post metamorphosis gas purging or, when useful, producing an artificial inner atmosphere, say of argon or helium, for the sample undergoing metamorphosis, otherwise when changing the inner working pressure both under vacuum and moderate over pressure. When the reactor works with liquid samples a suitable vessel is used for hosting the sample itself. The vessel is put inside the reactor core and again its shape, as for the sonotrode, is formed according to the angle law.

In the next paragraph, we report the results of the nuclear metamorphosis applied by the reactor to the element mercury.

18.3 Nuclear Metamorphosis Products

The reactor was firstly tested using a mole of mercury, 200 grams, which underwent nuclear metamorphosis during 6 min working time. The metamorphosis started after a very short inertial time interval, less than 10 seconds, producing the materialization of a dark ring of solid material around the sonotrode body on the mercury upper surface, as shown in Figs. 18.3–18.5. During the first tests the gases produced inside the atmospheric chamber were collected but not analyzed. The solid product material was analyzed thru many different

Fig. 18.3. Photo of the dark ring of solid material around the sonotrode head on the mercury upper surface during the metamorphosis.

Fig. 18.4. Samples of one mole of mercury. Left: after metamorphosis treatment Right: before metamorphosis.

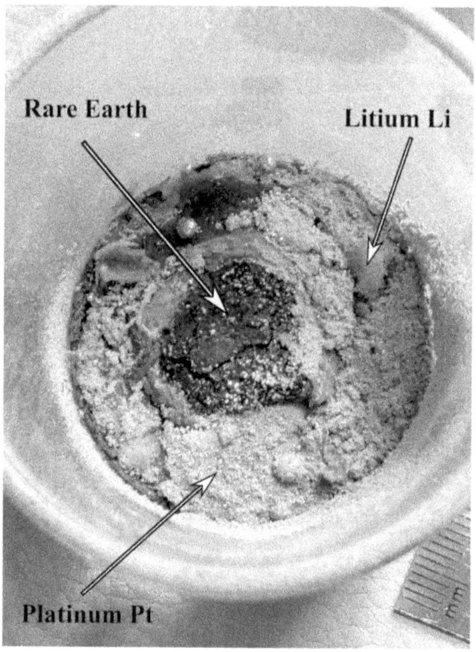

Fig. 18.5. Sample of Fig. 18.4 in detail.

techniques and the results are reported in Table 18.1, in the products were detected all the non gaseous elements of the periodic table. But in Table 18.1 all the elements already present in materials of the reactor and of the vessel are duly excluded [2]–[4].

Table 18.1. Reaction products of the DST reactions in mercury ordered by atomic number Z.

Z	Rare Earth	Element/isotope	Other detected element/isotope	Elemental concentration
3		^{7}Li(1st)		0.40 ± 0.05 ppb
4			^{9}Be (Mono)	0.004 ± 0.001 ppb
22		^{47}Ti (3rd)		7800 ± 800 ppm
23		^{51}V(1st)		0.10 ± 0.03 ppb
27		^{59}Co (Mono)		34 ± 2 ppm
28		^{58}Ni (1st)		186 ± 20 ppm
31		^{69}Ga (1st)		84 ± 10 ppm
		^{71}Ga (2nd)		
32			Ge	2780 ± 300 ppm
34		^{78}Se (2nd)		240 ± 25 ppm
		^{82}Se (4th)		
35		^{79}Br(1st)		77000 ± 5000 ppm
37			Rb	335 ± 35 ppm
39	Rare-earth		^{89}Y (Mono)	56 ± 8 ppm
42			^{95}Mo (3rd)	1113 ± 3 ppm
			^{98}Mo (1st)	
48			^{111}Cd (3rd)	6 ± 2 ppm
49			^{113}In (2nd)	10820 ± 120 ppm
50		^{118}Sn (2nd)		6 ± 0.6 ppm
		^{124}Sn (6th)		
51			^{121}Sb (1st)	0.30 ± 0.13 ppm
			^{123}Sb (2nd)	
58	Rare-earth	^{138}Ce (3rd)		34.5 ± 4.0 ppm
		^{140}Ce (1st)		
63	Rare-earth		^{151}Eu (2nd)	31 ± 7 ppb
64	Rare-earth		^{152}Gd (7th)	417 ± 17 ppm
			^{158}Gd (1st)	
70	Rare-earth		^{174}Yb (4th)	57 ± 24 ppm
71	Rare-earth		^{176}Lu (2nd)	141 ± 37 ppm
72		^{177}Hf (3rd)		450 ± 50 ppm
75			^{185}Re (3rd)	30 ± 3 ppm
78			^{190}Pt (6th)	1332 ± 9 ppm
			^{195}Pt (1st)	
			^{196}Pt (3rd)	
79		^{197}Au (Mono)		0.07 ± 0.01 ppb
90		^{232}Th (Mono)		35 ± 5 ppm
92		^{238}U (1st)		114 ± 20 ppm

Notes: In bracket, isotope position by natural abundance on the Earth: 1st, 2nd, 3rd, ... isotope; Mono = naturally monoisotopic element on the Earth.
The elemental concentrations in column 5 are related to the analyzed sample, not the whole mole of mercury metamorphosed.

The metamorphosis products reported in Table 18.1 are sketched in Fig. 18.6 along the curve of the nuclear binding energy per nucleon with respect to the atomic number, the nucleons of a nucleus. We see that in general the DST reactions are open in both directions starting from the mercury, towards nuclei lighter than mercury by nucleolysis and towards heavier ones by nucleo-synthesis. More in general these changes in the atomic weight without production of gamma radiation and radionuclides are made possible by the deformed space time reactions which, we stress again, give rise to the possibility that the course of the whole Aston-Bohr-Wheeler 'nuclear curve' is open in both directions.

The principal reasons to use mercury to test the reactor are the following. The nuclear energy density of mercury is about 8 MeV/nucleon, which is not far from the corresponding value in iron 8.9 MeV/nucleon. Thus, as verified, some minutes were deemed to be the necessary time to reach the DST reactions thresholds, both in space volume and in time interval. The mass density of mercury and

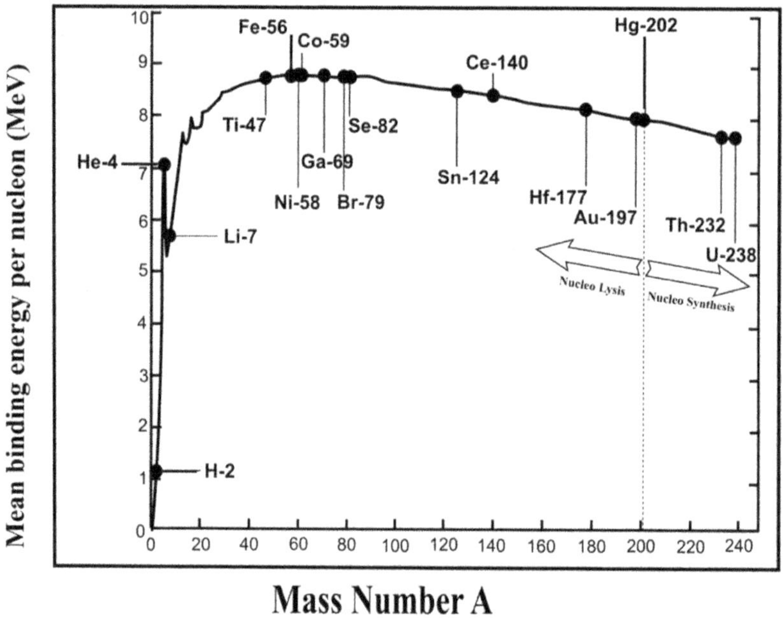

Fig. 18.6. Curve of the mean binding energy per nucleon E/A (MeV) with respect the atomic number A, the total number of nucleons in each nucleus. In black circles DST elements products, see Table 18.1.

its thermodynamic properties allow the supply of a power between 10 and 110 Watt/mole. This range was considered an 'energy rate', or better an 'energy density in time' suitable for violating the Lorentz invariance.

Moreover during the working of the RaM a macroscopic violation of the second law of thermodynamics seems to occur in the nuclear metamorphosis of the mercury. In fact, a liquid system receiving energy from an external device and increasing its temperature to about 240°C, is partially transformed into a solid state system, made of different elements, at a temperature more than 300°C higher than the solidification temperature of the starting liquid material. In other words, the degrees of freedom of the system decreased in the transition from liquid to solid, while its temperature increased. On the other hand, the second law of thermodynamics is not violated, if we take into account the microscopic non-thermodynamical nature of the occurring reactions, which are of nuclear type, not chemical. A nuclear source of free energy makes the systems an open system, while thermodynamics describes the properties of the entire ensemble not of a specific part. The Lorentz invariance violation, underlying at microscopic level this macroscopic 'metamorphosis' from liquid into solid, is a further reason for the apparent violation of the second law. In fact, in the LLI violation in electromagnetic systems the Clausius postulate for electric charge distribution was violated and thus the Clausius enunciation of the second law of thermodynamics was violated too.

Considering all the above facts, we remark once again that DST reactions are not 'nuclear transmutations', which correspond to transformations with radiations, but rather transformations of a new type, able to increase the atomic mass by nucleosynthesis or decrease it, by nucleolysis, in a very general way. It is well known that nuclei heavier than Iron can produce fission reactions thus producing lighter elements while those lighter then Iron can give rise to fusion reactions, thus producing heavier elements. Therefore, since in general changes of nuclear weight occur in both directions starting from a same nucleus, we introduce a new word for these changes occurring in DST condition: 'metabarysis'. It is formed by two terms from ancient Greek language: 'meta' (μετά) indicating change and 'barys' (βαρύς) indicating weightiness. By the way a reaction loop can be envisaged were the reaction products are the starting point for a process in the opposite direction. Thanks to the available reactor, in order to get a systematic use of the nuclear metamorphosis two ways can be followed. Either a very large number of tests are performed starting from different elements and their isotopes, thus obtaining a complete map of all the reaction, or general laws are found able to predict the possible, or at least the most probable, reactions products. The latter way is, from our point of view, more appealing, as it implies the knowledge of the underlying driving mechanism. The first step towards this latter direction, is to put the metamorphosis products in the plane N, Z, where the horizontal axe is N the number of neutrons and the vertical axe is Z the number of protons. Then we consider all the straight lines connecting each nucleus to each other. We warn that in general a sequence of nuclei along a single straight line does not necessary imply a sequence of alpha emissions from one to another nucleus. It rather corresponds to a parent nucleus that breaks into smaller parts and many of these parts are alpha particles. It is of interest to note, however, that we are trying to observe, from a not-deformed Minkowskian space-time, external to the nucleus, the consequences of reactions occurring in zones of deformed non-Minkowskian space time. A further parameter introduced by DST theory is energy. Thus, beside the two axes Z and N a third axis indicating the energy of the nucleus has to be considered. The occurrence of energy thresholds for DST reactions should correspond to particular surfaces in this three dimensional plot. In an attempt to predict the products of a DST reaction, the first step is to give an order to the products of the metamorphosis reaction. In N – Z (number of neutrons N – number of protons Z) plane we considered all the straight lines connecting each nuclide to the other. We expect the most of these lines to have a slope of 45°, corresponding to a difference of neutrons in the nucleus equal to the difference of protons, we remember that this angle is the first one coming from the Lorentz invariance violation forming the first part of the 'angle law'. All these segments at 45° do not lie on the same straight line, but rather in two parallels. The shorter distance between two nuclei belonging to the same straight-line corresponds to two neutrons and two protons, i.e. to an alpha particle. Moving downwards along a single straight-line does not necessary implies a sequence of alpha emissions from atom to atom. It rather corresponds to a parent nucleus that breaks into smaller parts and many of these parts are alpha particles. Following these assumptions, the occurrence of alpha-clusters is mainly a consequence to the curve of stable nuclides, as it is well known in the composite-nucleus model. However, it is worth strongly stressing once again that we are trying to observe from the point of view of a flat Minkowskian space-time, external to the nucleus, the consequences of reactions occurring in regions of deformed non-Minkowskian space time. Moreover, as already said, taking into account the fact that the determining phenomenological variable so that the DST reactions occur is the energy density, we introduce a third axis that considers the energy density per nucleon of the various elements.

In the plan (N), (Z), (E/A) we consider the curve formed by the nuclides: the 'nuclear plot'.

By considering equienergetic planes (E/A = const.) it is evident that elements with nuclei profoundly different have in fact the same energy density. This consideration can easily explain the reason why the 'DST nuclear methabarysis' where 'elements' of the nucleus which normally break down, under DST conditions can be joined (nucleo-synthesis) and those joining, instead can split (nucleo-lysis).

The well known nuclear plot in Fig. 18.7 refers to non DST conditions.

Surely we can consider that there is a plan equienergetic E/A which is the threshold triggering the DST reactions. Above this plane all the elements can be transformed in the same way: so actually the nuclear plot as a 'curve' no longer exists but it flattens out to a straight line. The distance between an element and its counterpart in the state DST is directly related to the energy that must be supplied

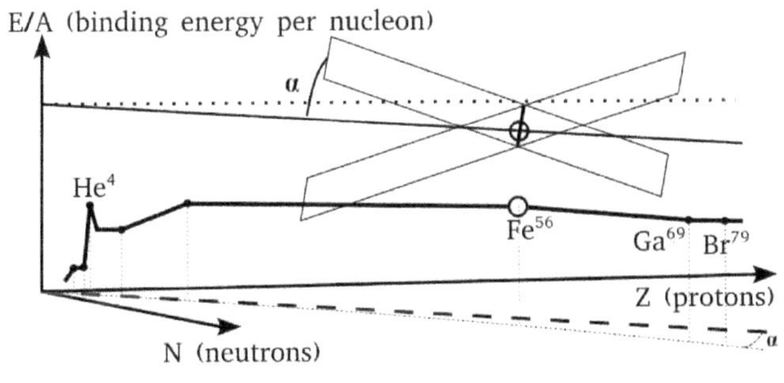

Fig. 18.7. DST transformation planes according the 'angle law', alpha represents the LLI violation angle.

to the system and it provides a direct measure. Our hypothesis is that each starting nuclide which exceeds the triggering threshold of the DST conditions will turn into the one/ones given by the intersection of the nuclear plot with the plane determined by (E/A) value of the nuclide in DST condition and having a precise angle of slope with respect to the equienergetic plane. The straight line fulcrum of inclination is determined by a law

$$Z = CN,$$

where $C =$ const., obviously with E/A = constant and equal to the level of the element into consideration conditions DST. As already shown the DST conditions determine by definition an alteration of the features of the transformation. At first analysis, therefore, we consider the plan of transformation as inclined with respect to the equienergetic one. Ceasing to supply energy from outside the DST conditions drop after a certain latency time, which means that there can not be only one plane of the transformation but also its mirror, which added up gives rise to a new equienergetic plane. In many cases both planes meet the curve of the elements under normal conditions in just two points thus resulting in the formation of the elements intersection, the nucleus and the nucleus formed by lysis or synthesis. In turn, these may exceed the threshold and thus give rise to other floors of the processing elements that will determine other products and so on up to the formation of elements not in DST conditions. Incidentally, by observing the curve of the elements in normal conditions, it is evident that those furthest from the DST threshold are lighter ones. The angles 'slope of transformation planes' are related to the 'angle law' for the violation of Lorentz invariance.

We will consider the evidences that angles of Lorentz violation should hold in an universal way and they could represent a fundamental asymmetry underlying all the natural phenomena both on a local terrestrial scale and on a cosmic one. Therefore it should be reasonable to do the 'ansatz' that such angles hold even on nuclear scale. The purpose is to apply the 'angle law' also to describe the DST transformations of the nuclei carried out macroscopically by nuclear metamorphosis, in order to make the DST reaction predictable and controllable [6, 7].

Of course there may be no single angle but, depending on the factors taken into account and the variation of their conditions, a law of variation of the angle is even possible. It also becomes clear that the identification of such a law would make the trends of the changes entirely predictable, making possible the realization of nuclear productions aimed at obtaining set of atomic weights of pre-fixed materials. As observed in all metamorphosis correctly performed in a complete way, the elements formed as a result of a DST reaction occur in their most stable isotope. This is due to the fact that their possible instability would provide energy to the DST metric, behaving as a sort of 'metric field', that acts so to maintain such a deformed condition, and then without emission of gamma rays. Indeed, evidence of such situation is the total absence of gamma rays in all the metamorphosis performed. As is known, the gamma rays are in fact energy and definitely help to keep the element under DST conditions. On the other hand, if the nuclear instability does not provide energy such as to maintain the DST conditions, then the nucleus 'falls' in flat Minkowskian space and therefore there is emission of gamma rays. Even the natural radioactivity, in normal flat space-time conditions, can behave in the same way, as in the experiment of radioactive nuclides we will treat in the next paragraph, they decayed much faster than in nature once it had been put in DST conditions. This means that it was providing energy to maintain itself within 'deformed' conditions inside the plane of the transformation, and therefore the transformation went on without any radiation emission or rather with reduced emission, say by providing themselves the energy that comes from their decay.

Neglecting now the detailed structure, we can therefore imagine the DST reactions as a 'transformer' between reacting elements and product elements and consider energy as a variable that intervenes in the sense of the 'generalized nuclear transformations': the nuclear metamorphosis.

And now comes the painful awareness. Nuclear metamorphosis may appear to be something like an 'arithmetic' of nuclei, simply the sum and subtraction of nuclei, with its lack of gamma radiation even casting doubt upon it. Tragically, we have no way to change this perception, but pointing to the experimental and phenomenological evidence well before us, Fig. 18.4 and Table 18.1.

We consider now the evolution of the deformed space time region which catalyzes the DST reaction and consequently the DST emissions. The DST emissions of neutrons and alpha particles together with the nuclei generated by nucleolysis and nucleosynthesis form the whole of the metamorphosis products. But the 'deformation' has its own life, or rather the energy 'stored' in the deformation follows both the DST reaction evolution and the metamorphosis evolution.

The metamorphosis evolution, as sum of the DST reactions, after the start of the supply of the critical energy density goes on after a delay of time, or an inertial time interval, before starting. Then it proceeds producing new nuclei and emissions. Lastly, when the supply ends, the system enters a latency time interval during which the production still lasts. For an example of this evolution see Fig. 6.5 sketching the neutron pulses in the time generated as DST emissions.

Then, at the end of each DST reaction, and therefore also when metamorphosis is finished, the energy stored in the deformation is yet extracted through the kinetic energy of the products. This kinetic energy is mainly carried by the lighter products, or rather neutrons and alpha particles, since the generated nuclei are slower and stable, moreover they are neither in excited state nor in metastable state, as consequence no gamma emissions were ever detected.

The kinetic energy of the products is mainly released within the vessel wall inside which the metamorphosis occurs, and it is detected as a huge increase of the temperature, fact verified in all the metamorphosis produced.

18.4 Nuclear Metamorphosis Neutralization

We now examine how the nuclear metamorphosis due to Deformed Space Time (DST) reactions on a large scale can produce the 'neutralization' of radioactive substances by a reactor using cavitation. The two cases examined are solution in water containing these radionuclides: the Thorium-228 and the Nickel-63.

About the thorium we have studied the hadro-leptonic thorium decay according to the Deformed Space Time (DST) theory.

The spontaneous decay of Th^{228} through the weak-leptonic interaction halves it in a time $t_{1/2} = 1.9\,\text{years} = 1.0 \times 10^6\,\text{min}$. The ratio between the half life of thorium, $t_{1/2}$, and the time interval of cavitation t_c to be used in the reactor, $t_c = 90\,\text{min}$, is $t_{1/2}/t_c = 10^4$. This means that cavitation brought about the transformation of Th^{228} at a rate 10^4 times faster than the natural leptonic decay would do. One is therefore led to deem that we are facing a phenomenon of metamorphosis more than an accelerated thorium decay, and that the metamorphosis of thorium into some other nuclide, induced by cavitation, is rather due to hadronic interaction, in particular to its non-Minkowskian part.

By bearing this in mind, it is possible to interpret $t_{1/2}/t_c = 10^4$ as the ratio between the decaying time of Th^{228} via the leptonic interaction (leptonic time t_{lep}), and the transformation time of Th^{228} via the hadronic interaction (hadronic time t_{had}). Namely, one has:

$$\frac{t_{1/2}}{t_c} = 10^4 = \frac{t_{\text{lep}}}{t_{\text{had}}}. \tag{18.4}$$

The Eq. (18.4) is the phenomenological basis from which we can start the theoretical deduction. Let us recall that the time elements of both metrics of electromagnetic and leptonic interactions, $b_{0,\text{e.m.}}$ and $b_{0,\text{lep}}$, are equal to each other, energy independent, and always equal to 1. Either metrics are therefore always Minkowskian with respect to the time. Moreover, the spatial elements of both metrics have the same functional dependence by the energy. Thanks to this circumstance, it will always be true, for the intervals of time $dt_{\text{e.m.}}$ and dt_{lep}, that $dt_{\text{e.m.}} = dt_{\text{lep}}$. Hence we can write:

$$\frac{t_{\text{lep}}}{t_{\text{had}}} = \frac{t_{\text{e.m.}}}{t_{\text{had}}} = 10^4. \tag{18.5}$$

On account of the hadronic law of time deformation, the same relation of proportionality holds between the threshold hadronic energy and the hadronic time and between the electromagnetic energy and the electromagnetic time. Therefore, it can be rewritten in terms of the time intervals $t_{\text{lep}} = t_{\text{e.m.}}$ and using the ratio of Eq. (18.4) as:

$$\frac{t_{\text{had}}}{t_{\text{lep}}} = \frac{E_{0,\text{had}}}{E_{\text{e.m.}}}. \tag{18.6}$$

From the earlier relation, it is possible to estimate the unknown variable $E_{\text{e.m.}}$, i.e. the energy transferred by the electrical (Minkowskian) interaction to the nuclei of thorium which get transformed into other nuclides by the hadronic interaction. One gets

$$E_{\text{e.m.}} = E_{0,\text{had}}\frac{t_{\text{e.m.}}}{t_{\text{had}}} = 367,5\,\text{GeV} \times 10^4 = 3.675 \times 10^{15}\,\text{eV}. \tag{18.7}$$

This value of energy is compatible with the maximum energy for nucleon which can be estimated for the metamorphosis induced by the cavitation. Let us recall that in order to get such an estimate, we have to proceed by considering power conservation, according to a continuity equation applied to the bubble collapse due to cavitation. The Eqs. (18.6) and (18.7) are the theoretical basis from DST theory to be used to tune and trigger the reactor for producing 'neutralization'.

A set of 12 identical solutions of Th^{228} in deionized bi-distilled water have been prepared, with concentration ranging from 0.01 to 0.03 ppb (part per billion). Th^{228} is an unstable element whose half life is $t_{1/2} = 1.9\,\text{years} = 1.0 \times 10^6\,\text{min}$. It decays by emitting 6 α and 3 β^-. The minimum emission energy of the alpha particles is 5.3 MeV, which is nearly equal to the energy of the α's emitted by the isotope of Radon, Rn^{222}. This alikeness allowed us to use the detector CR39, a polycarbonate whose energy calibration is specifically designed to detect those emitted by Rn^{222}.

Eight solutions out of the twelve were divided into two groups of four, and each of them was cavitated for $t_c = 90\,\text{min}$ at a frequency of 20 kHz and a power of 100 watt. The remaining four were not cavitated, and regarded as reference solutions.

The 12 detectors CR39 corresponding to the 12 solutions were examined and the traces impressed on them by the alpha particles counted.

The results obtained are depicted in Fig. 18.8. Precisely, the first column shows the four detectors CR39 used with the four noncavitated solutions taken as reference, whereas in the second and third columns one sees the eight detectors used with the eight cavitated solutions. The circles in the figure highlight the traces left by the particles α produced by thorium decay. Of the four CR39 used with the four reference solutions one counted 3 traces of alpha particles in all. The same number of traces was counted on the eight CR39 used with the eight cavitated solutions.

Then, the ratio of the number of traces and the number of solutions is 3/4 for the reference solutions and 3/8 for the cavitated ones. Thus, there is evidence of a reduction of the number of traces of alpha particles from thorium decay in the cavitated solutions with respect to those in the noncavitated ones. In particular, it is evident that the reduction in the number of traces from the former to the latter is by a factor 2.

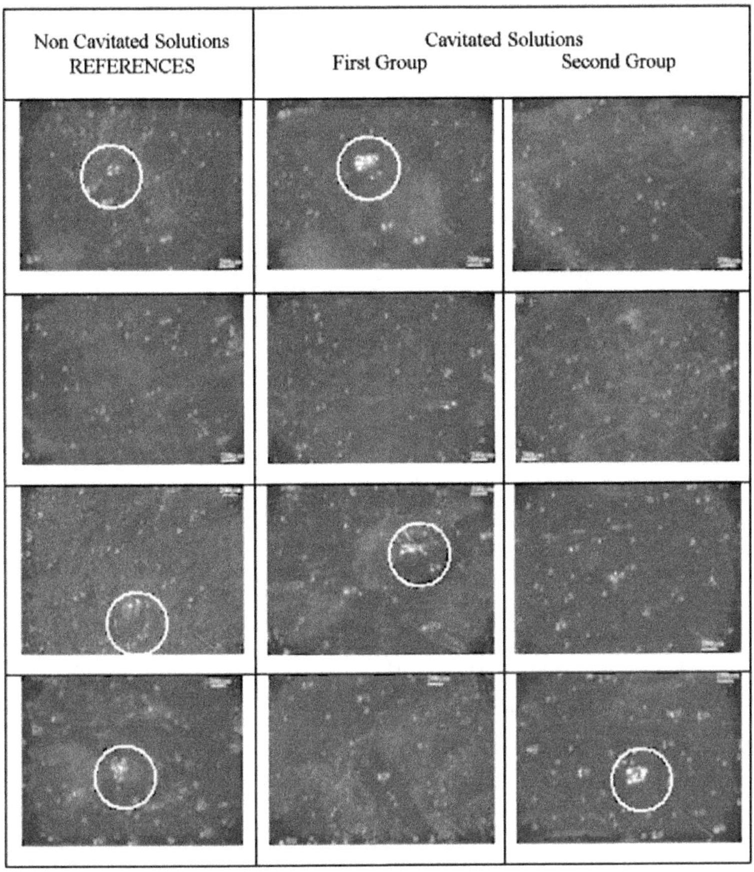

Fig. 18.8. Traces left by α-particles emitted from thorium decay on detectors CR39 (circles) for noncavitated (left) and cavitated (right) solutions.

The content of Th^{228} in those solutions providing evidence of alpha particles from thorium decay was analyzed by a mass spectrometer. The amount of Th^{228}, both in ppb and in counts per second, cps, found in the three cavitated solutions, whose CR39 showed the traces of the alpha particles emitted by thorium, is half of that in the three reference solutions corresponding to an α-emission. The results are reported in Tables 18.2 and 18.3, and allow to conclude that the process of metamorphosis reduced the content of Th^{228} in the solutions [5, 8].

Table 18.2. Content of Th^{228} in noncavitated reference solutions.

Mass-spectrometer analysis	Cps	ppb
Sample 1	287	0.020
Sample 3	167	0.012
Sample 4	363	0.026
Mean Values	272.33	0.019

Table 18.3. Content of Th^{228} in cavitated solutions.

Mass-spectrometer analysis	Cps	ppb
Sample 1 (first group)	231	0.016
Sample 3 (first group)	57	0.004
Sample 4 (second group)	79	0.006
Mean Values	122.33	0.009
Ratio of mean values noncavitated/cavitated	2.2	2.1

The next experiment to test the 'neutralization' by nuclear metamorphosis used a sample of nickel nitrate ($Ni(NO_3)_2$) made of radioactive Nickel-63, in nitric acid (HNO_3). This sample is more complex and more realistic than the pure thorium solution studied, as the radionuclide is a part of a molecule also containing other elements. This fact makes the conclusions of more general application. In addition, the sample is very corrosive, as it is the case of most nuclear waste, either of industrial or sanitary source, that are the most critical to be treated. Thus, the reported experiment can be considered also as a positive test for future applications in severe conditions, for instance in the case of radioactive and corrosive samples [9].

Radioactive Nickel-63 in nickel nitrate molecules dispersed in nitric acid, 67 %, 1.5 molar, were subjected to metamorphosis by means of a reactor at metamorphosis (RaM) but changing the frequency and the transferred power of the ultrasound applied. Nickel-63 is a beta-emitter, 66 keV being the maximum energy of the beta electron, with a half-life of 101.2 ± 1.5 years. The sample total mass was 39.6 g, the total volume was 30 cm^3: 25 cm^3 of nickel nitrate molecules in nitric acid and 5 cm^3 of bi-distilled and deionized water. The sample was sonicated in two runs of 100 seconds each. The sonotrode was the Aluminium alloy ERGAL 7075. The amplitude of the sonotrode tip oscillations was 7 μm with frequency 35 kHz and transferred power of 15 W. In the reactor the geometry of the sonotrode and sample vessel for sonication was as ever chosen in accordance with the parameters imposed by Lorentz violation in the framework of the DST theory, as done in former DST reactions: for instance, when nuclear metamorphosis occurred in mercury. Vessels of different shape and sonotrodes of different materials were used. The increase of temperature was assumed as an index to evaluate the efficiency of the whole apparatus vessel/sonotrode. When the highest temperature increase was obtained, the below reported measurements took place. A dosimeter-ratemeter (S.E. International, Mod. Radiation Alert), a Geiger counter (model Gamma Scout^{TM}) and a portable NaI spectrometer ATOMTEX (Model AT6102, n.61154) were used to detect the emitted radiation. They measured the total intensity or counts of beta, gamma and X-rays. The NaI spectrometer, placed at 0.5 cm apart from the sample, recorded the energy spectrum of the X-rays emitted by bremsstrahlung by the beta particles emitted by Ni-63. The actual measurement, aimed at detecting the reduced activity of the solution, was carried out by the NaI X-ray and gamma spectrometer. As a matter of fact, it was impossible to directly detect the beta electrons emitted by Ni-63, because even the most energetic ones at 66 keV were stopped by 1 cm of water, the thickness of the vessel, and the case of the NaI spectrometer. It is easily verifiable, for example through the application ESTAR of the US National Institute of Standards and Technologies (NIST), that electrons of 66 keV are completely stopped by 0.18 cm of glass and a bit more than 0.3 cm of water and of polyethylene which may simulate just the external case of the NaI spectrometer. In this sense, the spectra shown in the following figures can be only interpreted as the bremsstrahlung X-rays of the beta electron of Ni-63, or rather X emission with a maximum of energy at 66 keV, the end point. The dosimeter monitored only the dose and dose-rate in the laboratory. The Geiger counter, on the other hand, having the possibility to measure without any screen but the thin beryllium window, was used to monitor the possible contamination of the surface surrounding the vessel and of the surface of the sonotrode. No significant variation with respect to the background in any part of the experiment was registered by any of these two instruments. For this reason, their measurements will not be reported. In Table 18.4, we report the different working parameters of the reactor in order to show that when the reactor correctly operates according the DST rules, say angle law, it is possible to work even at different conditions obtaining the same result of nuclear metamorphosis of treated substance.

Now let us consider first the differences between the two experiments, as reported in Table 18.4. The geometry of both the sonotrodes was chosen by following the rules considered mandatory for inducing the violation of Local Lorentz Invariance and hence the Space-Time deformation. The sample's container, on the contrary, was chosen without following this kind of rules in the case of thorium, while, for Nickel, it is a vessel with an appropriate geometry. These geometric rules aim to obtain an appropriate distribution of energy concentration. In fact, the DST nuclear reactions were observed to occur or not to occur depending on the energy distribution in the previous experiments with Mercury. Both for Thorium and for Nickel, the micro-reactors are the bubbles of appropriate size contained in the solution and made to collapse by sonication and cavitation. In conclusion, the main difference between the two experiments consists in the geometry of the container: the vessel rather than the reaction chamber. In terms of acquired know-how, this means that the reactions in thorium were produced just by choosing the appropriate bubble size, while the chamber geometry design converged on the right one for well-working bubbles by sonication, without taking

Table 18.4. Experimental details in thorium experiment and in Nickel experiment.

	Thorium experiment	Nickel experiment
Ultrasound frequency	20 kHz	35 kHz
Ultrasound transferred power	130 W	15 W
Amplitude of the sonotrode tip	30 microns	7 microns
Sample mass	300 g	39.6 g
Sample volume	300 cm^3	30 cm^3
Sonication runs	1 run of 90 min	2 runs × 100 s
Sonotrode composition	Steel alloy	Aluminium alloy
Sample container	Reaction chamber	Vessel

other suggestions from DST theory into account. After the thorium experiment, we chose the vessel geometry to obtain the proper energy distribution or 'densification' by following the asymmetry of Lorentz violation, as resumed in the 'angle law', in order to enhance the whole effect of collective DST reaction up to the macroscopic level of the nuclear metamorphosis, even reducing the time in the Eq. (18.5).

Therefore, we used this new vessel for the nickel experiment, quite different from the reaction chamber. In this way, we exploited all the conditions for the DST-reactions: space, time and asymmetry.

From a technological point of view, the two experiments were built around two different types of working conditions of the reactor, which has been varied in the cooling device. In the first one on thorium, the reactor was equipped with an effective cooling system which allowed an uninterrupted sonication for 90 min. In the second experiment performed on nickel the reactor cooling system was turned off, limiting the maximum duration of a run to 150 s and operationally fixed at 100 s for precaution. The over-heating due to longer sonication would have damaged the instrument: the sonotrode would have changed its length and the resonance conditions therefore the right working parameters, employed to have the adequate energy distributions, would have been lost. Additionally in order to preserve the integrity of the system, the ultrasound machine was implemented with a security system blocking the electrical supply, thus stopping the machine, if working out of the right parameters.

Notwithstanding the difference between the two experiments, the condition of local Lorentz invariance (LLI) violation was achieved in both cases and the 'neutralization' took place.

In the Nickel experiment, lower power and shorter run-times could be used thanks to the use of the knowledge of the 'angle law' which once more unveils its double role of both a law of nature and a technological rule.

In this way, the right energy densities, in space, time and direction, have been reached at the same time to trigger the DST reactions to a large scale involving a number of nuclei comparable to Avogadro number. This affects the bulk of the substance which gives rise to the macroscopic effect of the metamorphosis, which in the case of radioactive substances produces the reduction of activity.

The measurements reported in Table 18.5 give 13% decrease of the emitted intensity which corresponds to a time lapse of about 20 years,

Table 18.5. The per-cent decreases of counts correspond to about 20 years of natural decay, as shown in Fig. 18.11.

Counts before sonication	2380 ± 110	% decrease
Counts after 100 s of sonication	2070 ± 110	$13 \pm 6\%$
Counts after 200 s of sonication	2046 ± 110	$14 \pm 6\%$

Note: Total counts of the bremsstrahlung radiation in 100 seconds, after background subtraction in the range: 0 to 66 keV): before cavitation, after 100 seconds of sonication and after 200 seconds of sonication.

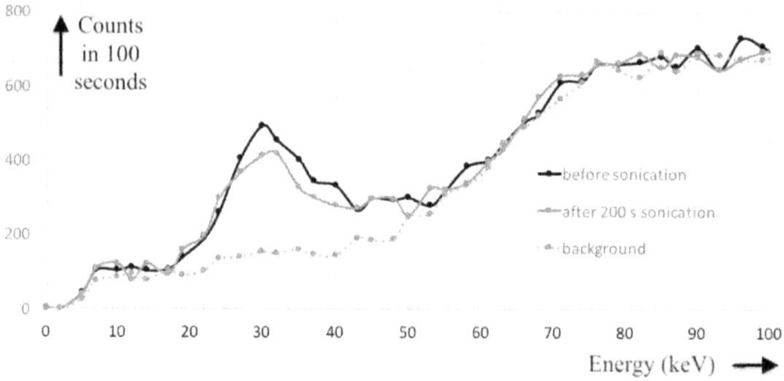

Fig. 18.9. Detected intensity (NaI portable X-gamma spectrometer ATOMTEX AT6102). Dotted line: Laboratory background, Black line: Ni-63 before sonication, Grey line: Ni-63 after 200 seconds of sonication.

as the decay constant of Ni-63 is 146.00 ± 2.16 years. An additional sonication of 100 s corresponded to a smaller reduction, the total decrease being $14 \pm 6\%$ after the two treatments. The smaller contribution of the second run was attributed either to the reduced number of bubbles suitable to act as micro-reactors of the DST reactions or to the occurrence of secondary competitor reactions that followed the primary reactions, thus absorbing energy. The detected background spectrum of the laboratory is shown in Fig. 18.9 as a dotted line. The emission of Ni-63 falls between 0 and 66 keV, a zone characterized by a low and almost flat background. Therefore, the bremsstrahlung radiation of the emitted beta-electrons also falls in this zone, as it is visible in Fig. 18.9 the continuous black line corresponds to the bremsstrahlung peak superimposed on the background. We remark that this peak does not correspond to the so-called 'inner bremsstrahlung' but to the spectrum detected after the absorption of the beta electrons partially inside the materials of the sample and the walls of sample-container. Due to this fact, the peak was well visible despite the moderate activity of the radioactive source. Thus, we avoided many problems of safety, transport and shielding and it was possible to carry out measurements with easily available standard instrumentation.

After a sonication aiming to induce the DST conditions, if the decay process was enhanced in such a way to reduce the time constant, an increase of the emitted intensity was expected. On the contrary, we stress that the already quoted detectors, dosimeter-ratemeter, S.E. International, Mod. Radiation Alert, and a Geiger counter, model Gamma Scout^{TM}, which recorded the radioactivity all along the experiment, did not register any detectable variation of intensity.

After each sonication session, the spectrometric measurements were repeated. The grey line in Fig. 18.10 reports the spectrum obtained after the second session, which corresponds to a total sonication of 200 seconds. A general decrease of the counts is visible. It is better to put into evidence in Fig. 18.10, where the counts corresponding to the emission energy of Ni-63 (0–66 keV) are reported after background subtraction.

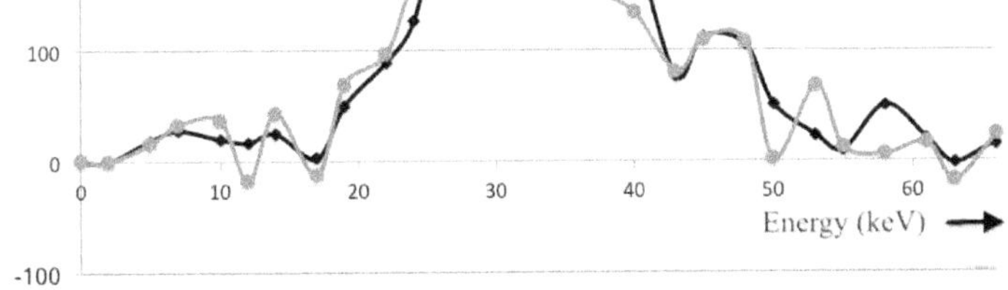

Fig. 18.10. Detected intensity from the bremsstrahlung radiation after background subtraction (region 0-66 keV) Black line: Ni-63 before sonication, Grey line: Ni-63 after 200 seconds of sonication.

This decrease was considered as the consequence of nuclear metamorphosis transforming the radioactive nuclei of Ni-63 into different nuclei. It is self-evident that a radioactive source with less emitting nuclei is less intense.

A quantitative measure of the effect can be found in Table 18.5, where the total counts of the region 0–66 keV are reported after background subtraction.

The measurements reported in Table 18.5 show a 13% decrease of the emitted intensity corresponding to a time lapse of about 20 years, as the decay constant of Ni-63 is 146.00±2.16 years. An additional sonication of 100 seconds corresponded to a smaller reduction, the total decrease being 14±6% after the two treatments. The smaller contribution of the second run was attributed either to the reduced number of bubbles suitable to act as micro-reactors of the DST reactions or to the occurrence of secondary competitor reactions that followed the primary reactions, thus absorbing energy.

The above reported considerations and experimental facts could easily lead to the seeming hypothesis that the decay constant is altered after the ultrasound application.

On the contrary, any possible deviation from the exponential law could be the result of two combined actions: DST transformations at the beginning of the treatment and return to the normal decay in the following phase.

The behaviour is depicted in Fig. 18.11, where the usual decay of Ni-63 is reported as a continuous line. The dashed curve corresponds to the sudden decrease, indicated as STEP, of the radionuclides in the first 100–200 s which is followed by a decrease with the usual decay constant. This way, the time necessary to reach 50% of the initial activity, which is usually indicated as the first half-life, is smaller by about 20 years as indicated by the two large circles in Fig. 18.11.

Thus, a decrease is obtained in the first phase of 200 s which should correspond to about twenty years of natural decay, as indicated by the small grey circle in Fig. 18.11. The result is quite noteworthy: in less than 5 min the half-life was shortened by about 20 years although the decay constant is the same [10].

We now report the results from mass spectrometry. To check the radiation spectrometry results, in particular that the number of Ni-63 nuclei decreases and they transform into different nuclei, measurements of inductively coupled plasma mass spectrometry (ICP-MS)

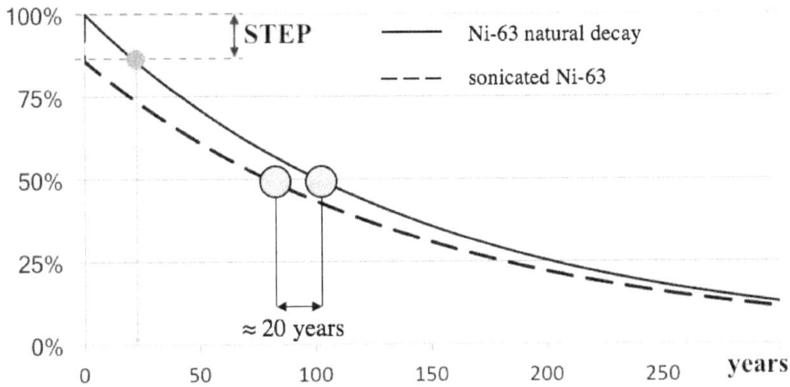

Fig. 18.11. Decay of Ni-63: Continuous line: the natural decay of the emitted intensity. Dashed line: the proposed decay after sonication: after the abrupt step in the first 200 s, the emitted intensity decreases by following the natural decay constant.

were performed by using an Agilent 7000 ICP-MS spectrometer. The studied sample was that from radiation spectrometry, which was sonicated 200 s. A reference sample, or 'blank', was obviously obtained without the final sonication.

The results obtained are shown in Table 18.6, where the percentage differences between the sample treated with ultrasound and the untreated one are reported. This percentage is defined as: 100% (S-B)/S where S is the content of the element/isotope in the sonicated sample and B is in the content of the element/isotope in the reference blank.

A first indication coming from Table 18.6 concerns the decreased number of Ni-63 nuclei: the 13±10% decrease obtained by ICP-MS technique, evaluated with respect to the sample, agrees with the 14±6% evaluated with respect to the blank, of the radiation spectrometry.

We remark that a decreased number of radionuclides and of the related emission, alpha emissions in that case, was also measured after metamorphosis of Th-228.

A second indication is that the concentration of other nuclei is altered as we can see from the results reported in Table 18.6.

In the sonicated sample, it is evident an increase with respect to the blank of Co-59, a stable isotope which can be supposed to be the

Table 18.6. Agilent 7000 ICP-MS analysis.

Element	Z	N	A	Sample – Blank Difference %	RSD	Notes
Li	3	4	7	35	±10	
Be	4	5	9	9	±10	
B	5	6	11	9	±10	
Na	11	12	23	94	±20	
Mg	12	12	24	90	±10	Ergal 7075
Al	13	14	27	95	±10	Ergal 7075
K	19	20	39	20	±10	
Ca	20	24	44	5	±20	
Sc	21	24	45			Not detected
Ti	22	25	47	97	±10	Ergal 7075
V	23	28	51	94	±10	
Cr	24	28	52	28	±10	Ergal 7075
Mn	25	30	55	25	±10	Ergal 7075
Fe	26	30	56	5	±10	Ergal 7075
Co	27	32	59	14	±10	
Ni	28	32	60	22	±10	
Ni	28	35	63	−13	±10	Ergal 7075
Cu	29	36	65	27	±10	Ergal 7075
Zn	30	36	66	40		
Ga	31	38	69	67	±10	
As	33	42	75	1	±10	
Se	34	43	77	16	±10	
Rb	37	48	85	29	±10	
Sr	38	50	88	33	±10	
Zr	40	50	90	73	±10	Ergal 7075
Nb	41	52	93			Not detected
Mo	42	53	95	10	±10	
Ru	44	57	101			Not detected
Rh	45	58	103			Not detected
Ag	47	60	107	−161	±10	
Cd	48	63	111	0	±10	
In	49	66	115	16	±10	
Sn	50	68	118	14	±10	
Sb	51	70	121	31	±10	
Cs	55	78	133	7	±10	
Ba	56	81	137	26	±10	
La	57	82	139	52	±10	
Ce	58	82	140	57	±10	

Note: Z: Atomic number; N: Neutron number; A: Mass number. In the column Notes, the elements/isotopes that are present in the sonotrode material (Ergal 7075).

(*Continued*)

Table 18.6. (*Continued*).

Element	Z	N	A	Sample – Blank Difference %	RSD	Notes
Rare Earths						Not detected
Hf	72	106	178		±10	Not detected
Ta	73	108	181		±10	Not detected
Pt	78	117	195			Not detected
Au	79	118	197			Not detected
Tl	81	124	205	35	±10	
Pb	82	126	206	15	±10	
Pb	*82*	*126*	*208*	*11*	*±10*	
Bi	83	126	209	5	±10	
U	92	146	238	23	±10	

Note: Z: Atomic number; N: Neutron number; A: Mass number. In the column Notes, the elements/isotopes that are present in the sonotrode material (Ergal 7075).

main candidate of the DST reaction of Ni-63 during the metamorphosis. The corresponding proposed reaction is:

$$\begin{array}{ccccc} {}^{63}\mathrm{Ni}^{28}_{35} & \rightarrow & {}^{59}Co^{27}_{32} & + & {}^{2}H^{1}_{1} + 2n \\ \text{decreased} & & \text{increased} & & \text{not detectable} \end{array}$$

A second hypothesis, based on the increase of Li-7, Be-9, B-11 and Ne-22, can justify the decrease of Ni-63 nuclei:

$$\begin{array}{ccccc} {}^{63}\mathrm{Ni}^{28}_{35} & \rightarrow & 3{}^{7}Li^{3}_{4} & + & {}^{9}Be^{4}_{5} \\ \text{decreased} & & \text{increased} & & \text{increased} \end{array}$$

$$\begin{array}{cccc} + & {}^{11}B^{5}_{6} & + & {}^{22}Ne^{10}_{12} \\ & \text{not detectable} & & \text{not detectable} \end{array}$$

These proposed reactions, triggered by the present knowledge, do not exhaust the scenario of all possible patterns opened by the nuclear metamorphosis. Mass spectrometry analysis and radiometric spectrometry results are in agreement, thus giving a solid basis to the neutralization of radioactive materials by nuclear metamorphosis [11]. It can also be noted that the elements produced by such DST reactions are not excited, as no gamma radiation was detected during or after the process of metamorphosis. According to DST theory, the energy that usually is taken away by the gamma radiation, in these cases is used to deform the space-time. In this case the Eqs. (18.5), (18.6) have been used for an 'a posteriori' theoretical test of the right use of the 15 watt transmitted power as tuning value for the reactor working conditions, since among the other conditions it was mandatory to take care also of the hadronic metric fifth element functional behaviour, as we will see in the next Section 18.5.

18.5 Test of the Hadronic Fifth Metric Element by Nuclear Metamorphosis

At last it was performed a test by the metamorphosis reactor in order to establish if the nuclear metamorphosis occurs following the fifth element of the hadronic metric which behaves as the sixth power of the interaction energy, as calculated by Ricci flow in Chapter 5. This power dependence can be effectively tested when the operational parameter of the reactor are well established and the metamorphosis occurs, any small change in their values can amplify or press down the phenomenon. In fact, as already said, a change of a factor ten can produce an effect of a factor one million, ten to sixth power. Obviously, for safety reasons, the chosen way was to change the parameters for obtaining null metamorphosis with respect to the optimal configuration. Therefore even keeping all the other parameters including the vessel and the mass of mercury, one mole, it was decided to maintain the shape of the sonotrode but to change symmetrically its profile of a small amount of at most 1/10, so it was expected that the metamorphosis process became depressed of a factor one million. The test occurred successfully and the mercury mole produced no metamorphosis products at all even if the reactor ran for 30 minutes, an interval at least five times greater than the 5–6 minutes needed in standard conditions to get metamorphosis. Besides the check of the dependence from the sixth power of energy according the hadronic fifth metric element, this test warns about how narrow the gauge ruling the nuclear metamorphosis is, and how delicate the tuning of all the parts and related working parameters of the reactor is.

References

[1] D. Bassani and F. Cardone, Sidom Technical Report 1-015, 2015.
[2] F. Cardone *et al. Int. Jour. Mod. Phys. B*, 29, 1550239-1/13, 2015. DOI: 10.1142/S0217979215502392.
[3] D. Bassani *et al. Int. Jour. Mod. Phys. B*, 13, 1750168-1/20, 2017. DOI: 10.1142/S0217979217501685.
[4] A. Rosada *et al. J. Condensed Matter Nucl. Sci.*, 27, 1–9, 2018.
[5] F. Cardone *et al. Phys. Lett. A*, 372, 22, 1956, 2009.
[6] R. Capotosto and F. Rosetto. *J. Adv. Phys.*, 5, 80–83, 2016.
[7] G. Albertini and R. Capotosto. *J. Adv. Phys.*, 5, 84–89, 2016.
[8] G. Cherubini *et al. Radiochimica Acta*, 107, 6, 469, 2019. DOI: 10.1515/ract-2018-3009.
[9] P. Avino *et al. Springer Nature Appl. Sci.*, 1, 1319, 2019. DOI: 10.1007/s42452-019-1391-6
[10] A. Rosada *et al. Int. Jour. Mod. Phys. B*, 34, 4, 2050001, 2020. DOI: 10.1142/S0217979220500010.
[11] E. Guerriero *et al. Int. Jour. Mod. Phys. B*, 35, 2, 2130001, 2020. DOI: 10.1142/S0217979221300012.

Part V

Cosmological Consequences

Chapter 19

Lorentz Violation and Neutron Emission

In Rome, it was carried out the experiment to study the integral distribution in space of the neutron energy produced by means of DST reactions. In previous experiments ran in Milan and Turin there was found DST neutron emissions from steel, ferritic iron, and basalt subjected to ultrasound [1, 2]. In the Rome experiment, a steel bar was employed, by subjecting it to ultrasound, and the produced neutrons were measured by means of polycarbonate PADC detectors in order to get both an integral and a polar measure of the neutron energies.

The experimental tests have been performed by exploiting the DST reactor, designed and built by the Italian Army in collaboration with Startec Ltd., Italian company operating in the field of ultrasonic machines, see Fig. 19.1.

The whole equipment was constituted by stable and supporting elements having as their objective to ensure, during the sonication, the system stability and immobility, as well as a complete and almost perfect contact between the plane surface of the sonotrode tip and the head surface of an AISI 304 cylinder. The piezoelectric column was connected to the generator, operating at the maximum power of 2 kW and at frequency of 20 kHz, by means of high frequency cables. Each run of the experiment was performed for 3 min of duration. All the sonication system has been cooled using a compressed air jet generated by a Vortex mod. 610 OSHA compressor.

The cylindrical bar, made by steel AISI 304, had the following dimensions: height 90.5 mm, diameter 18.8 mm, weight 181 g.

The composition of this bar has been measured by X rays fluorescence (XRF) performed on the external surface: Fe 69.99 %, Cr 18.57%, Ni 8.71%, Mn 2.20%, Mo 0.32%, Ti 0.109%, V 0.077%, Cu 0.034 %. The AISI 304 cylinder has been inserted inside a supporting cylindrical block made by PTFE, Teflon, 7 cm high, 6 cm diameter, see Fig. 19.1. The Teflon cylinder containing the steel bar has been set vertically inside the Neoprene frame and has been fixed on the base of the reactor. Two neutron detectors were placed on the same plane parallel to each other, in front of the system composed of AISI 304 steel cylinder — PTFE support and 30 cm away, see Fig. 19.1. The first one, a Microspec Neutron Probe by BTI, measured the neutron emission, recording their energy, the flux, and the dose. The second one, the HDS 101 by Mirion, for the neutron count rate and for the dose measurement, was equipped with a Li I(Eu) neutron detector, as well as with a Cs I(Tl) scintillation detector for γ rays measurement.

The Teflon cylinder goal was to act as a calorimeter; for this reason, hereafter, we will refer to it as 'Teflon Calorimeter', or in brief calorimeter. It was externally surrounded, to about half of its height, by a circular array of sixteen passive neutron detectors model CR39-AB. They are PADC detectors corrected with boric acid, whence the AB name, see Fig. 19.2. Hereafter, for brevity, we shall refer to them simply as CR39, unless differently need. Each CR39 platelet was placed with its surface perpendicular to the horizontal

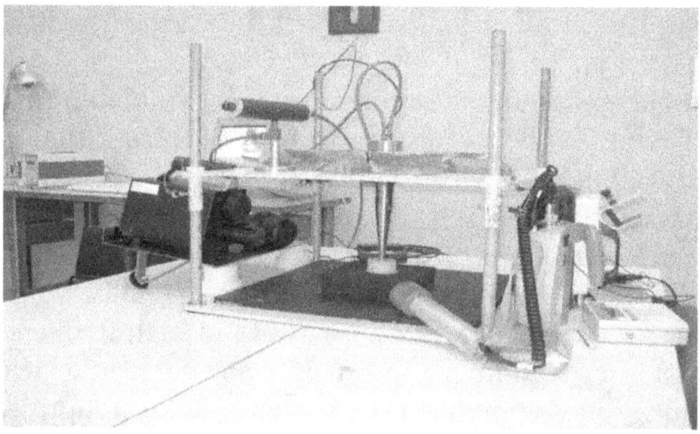

Fig. 19.1. General view of the Italian Army (E.I.) DST reactor used. For kind permission by the State Undersecretary of Defense and the Head of Staff of Defense, in the years 2010–2011. Subsequently it has been dismantled.

Fig. 19.2. Close-up of the experimental devices details, the sonotrode, the cooling rings and the Teflon calorimeter surrounded by CR39-AB.

plane and its face parallel to the vertical one. The laboratory background was measured many times, both with HDS Mirion and with BTI Microspec, as well as with four CR39 detectors, placed in the same laboratory where the experiment was run, at a distance of 20 m from the equipment. The time of each experimental run of steel rod sonication was of 3 min. The HDS Mirion and BTI Microspec measurements have been acquired, stored and computed by means of an Olivetti mod. 'Punto it' computer. At the end of the experimental run the CR39 detectors were picked up, and observed and photographed by means of an optical microscope 100× and an electronic microscope.

In a previous and separate phase of the experiment, all the three detector types were calibrated by an Am-Be source, as well as by means of the TRIGA Mark II and TAPIRO nuclear reactors of the ENEA Research Center at Rome-Cesano, by exposing them, for 15 min, at a neutron flux of $1.3 \times 10^{10} \mathrm{n} \times \mathrm{cm}^{-2} \times \mathrm{s}^{-1}$ and $4 \times 10^{12} \mathrm{n} \times \mathrm{cm}^{-2} \times \mathrm{s}^{-1}$, coming, respectively, from the horizontal and the radial channel of the thermal column. The photographic technique on neutrons filtered into boric acid dust was independently developed, since 2006, at the ENEA Research Center in collaboration with CNR-ISMN, the photographic technique of neutron measurement with boric acid was further developed and definitively tested as safe and reliable [3].

The measurement synoptic summary is shown in Fig. 19.3, where the steel bar, inserted in the Teflon, is displayed in top plane view. We stress that the HDS Mirion detector did not record any γ-ray emission higher than the environmental background, moreover, it did not detect neutron emission. On the contrary, the BTI Microspec detector measured an emission characterized by a 110 pulses spike, at the maximum peak height, corresponding to a neutron energy of about 0.7 MeV. The whole dose measured was 0.024 μSv.

The background measurement, lasted 15 min, resulted in a dose of 0.003 μSv, it is reported in the bottom left frame of Fig. 19.3. The CR39 detectors were taken from their positions around the Teflon cylinder and were subjected to a specific treatment, namely to highlight the neutron traces by soaking the film for 4 hours in a NaOH thermostatic bath at the controlled temperature of 82°C. After the treatment, the CR39s were dried at ambient temperature. The so obtained images on CR39 were observed and photographed by an optical microscope at 100×. On the detector photos, squares of 10 cm side, the neutron trace counting was performed by subdividing each photographic image in 400 equals parts by means of a grid of small squares of 0.5 cm side.

Identical procedure was exploited for the four blank CR39 detectors, as well as for the reference images from detectors of the same type subjected to neutron irradiation in the horizontal channel of the thermal column of the TRIGA reactor and the radial 2 channel of the TAPIRO reactor, respectively. For the visual counting of the neutron traces in the CR39s, for each pane of the grid a yes/no criterion was adopted: emission trace=yes, trace absence=no. The results are shown in Fig. 19.3, where, in horizontal view, the disposal of the CR39s, each denoted by its respective serial number, and the HDS Mirion and BTI Microspec detectors layout, are sketched with respect to the steel bar and to the Teflon calorimeter. The general plant is subdivided in four quadrants of 90° of amplitude, each comprising four CR39s, with an orientation South-North, according the convention ever used in the angle law experiments of magnetic coil

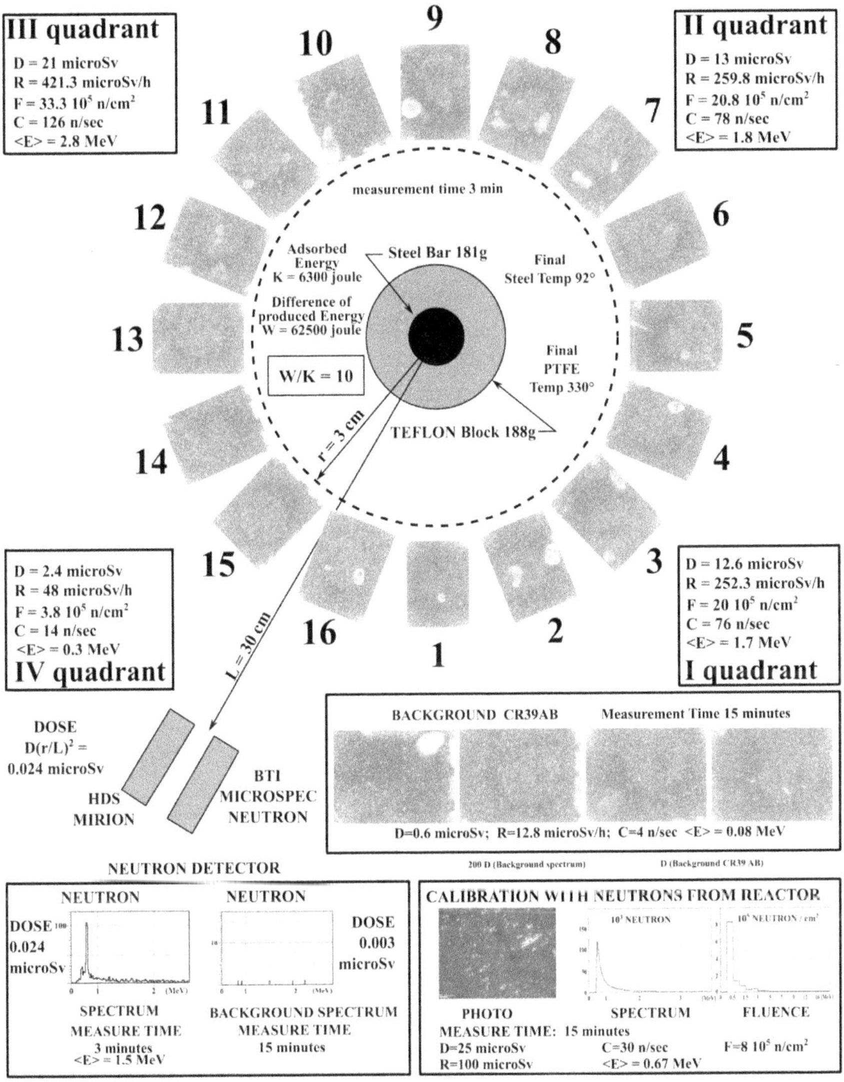

Fig. 19.3. Sketch of the DST reactor horizontal view, showing the AISI 304 steel bar inserted in and surrounded by PTFE cylinder, and synoptic summary of measurements. Externally, by moving anticlockwise from n.1 to n.16, the icons of the CR39-AB detectors with the traces of the alpha particles generated by the neutron impact with the Boron dust covering the CR39 plate. Note that the ratio between adsorbed energy K and the produced energy W is W/K = 10.

and torsional antenna. Moving anticlockwise, we get the following results. In the first quadrant, a 12.6 μSv dose was measured, corresponding to a dose rate of 253.3 μSv/h, and to an integrated neutron flux, 3 minutes measurement, of 2×10^6n $\times$ cm^{-2}, equivalent to 76 n $\times$ s^{-1}, with the maximum neutron energy of 1.7 MeV. The integrated flux was calculated by direct proportion from the calibration executed by means of the corresponding TAPIRO reactor neutron flux. In correspondence of the second quadrant, a 13 μSv dose was measured, corresponding to a dose rate of 259.8 μSv/h, and to an integrated neutron flux of 2.08×10^6n $\times$ cm^{-2}, equivalent to 78 n $\times$ s^{-1}, the maximum neutron energy was 1.8 MeV. In the third quadrant a 21 μSv dose was measured, corresponding to a dose rate of 421.3 μSv/h, and to an integrated neutron flux of 3.33$\times$106n $\times$ cm^{-2}, equivalent to 126 n$\times$s^{-1}, the maximum neutron energy was 2.8 MeV. In the fourth quadrant, a 2.4 μSv dose was measured, corresponding to a dose rate of 48 μSv/h, and to an integrated neutron flux of 3.8 $\times$ 105n $\times$ cm^{-2}, equivalent to 14 n $\times$s^{-1}, the maximum neutron energy was 0.3 MeV.

For the whole 360° angle the total dose measured by means of the CR39 was 49 μSv, corresponding to a dose rate of 981 μSv/h, and to an integrated neutron flux of 5.8 $\times$ 106n $\times$ cm^{-2}, equivalent to 294n$\times$s^{-1}, the maximum neutron energy was 2.8 MeV. The concomitant absence of counts for the HDS Mirion detector can be explained by the fact that it converts, as the majority of the active detectors, the neutron energy in an electrical signal, for this reason, because of the dead time due to the electrical conversion of the pulses, it may have failed the detection. Moreover, a further explanation is provided by the dramatic decrease of the detector efficiency with the square of the distance from the source. The law $1/R^2$ can justify the difference between the neutron measurements in the fourth quadrant obtained by means of the BTI Microspec detector and those by the CR39s. Note that, while the distance of the CR39s from the AISI 304 steel bar was 3 cm, the Teflon calorimeter radius, the distance of both the active detectors was 30 cm, as can be seen in Figs. 19.2 and 19.3. This fault of the active neutron detectors, compared to the passive ones, is expected to be emphasized when the neutron emissions are not constant over time and occur by means of neutron quick bursts, as reported in Chapter 6. The mutual temporal distance and amplitude of such bursts are knowable with difficulty and then, at first sight, random, as already discussed in Chapter 6. After dividing the CR39 measurements in quadrants, the individual detector counts were analyzed by keeping a strictly two-sided orientation, going on counterclockwise, fixed on hereafter as a conventional way. The results are shown in Figs. 19.3 and 19.4(b). For each detector, in addition to the serial number, the dose in units of μSv and the angle measured with respect to the conventional sided axis are reported. In Fig. 19.4(b), the length of the bold lines, starting from the center in the radial direction, provides a visual description of the measured dose, resulting in a true polar histogram of the measured energy carried by the emitted neutrons.

This dose includes also the energy absorbed by the Teflon calorimeter, that reached the fusion temperature, $T_{\text{Teflon}} = 327°\text{C}$. The energy has been calculated, in direct proportion with the energy corresponding to each CR39 detector measurement, after the calibration performed by the radial channel 2 of the nuclear reactor TAPIRO, see Fig. 19.3, right bottom frame. By going around the picture by steps of 22.5°, a clear asymmetrical distribution of neutron pulses is evidenced, as already envisaged at the end of Chapter 15 as analogy with the torsional antenna polar diagrams of radiant energy. The emission clusters reach the maximum, corresponding to a total dose of 110 μSv, within an angle of 90° between 157.5° and 247.5°. Inside this arc, we clearly see that the emission looks isotropic, but outside it the emission is no longer such. On the contrary at the 90° angle between 270° and 360°, the emission is minimal corresponding to a total dose of 16 μSv.

Moreover, it is quite evident, again from Fig. 19.4, that there are two mutually orthogonal directions having these special features: along one the dose is null, along the other one the emission has the maximum dose. With reference to Fig. 19.4, these two directions are evidenced according to the convention adopted in Fig. 19.3: the first marked by number 10 upside (3rd quadrant) and number 2 downside (1st quadrant), the other one marked by number 6 upside (2nd quadrant) and by number 14 downside (4th quadrant). We stress again that the neutron released energy, the emission dose, is evidently anisotropic. In fact, the emission intensifies just in correspondence of the 157.5–247.5° angle. Such asymmetry and anisotropy of the neutron energy emission give us the evidence that we are in presence of reactions that occur according both to the formalism of the

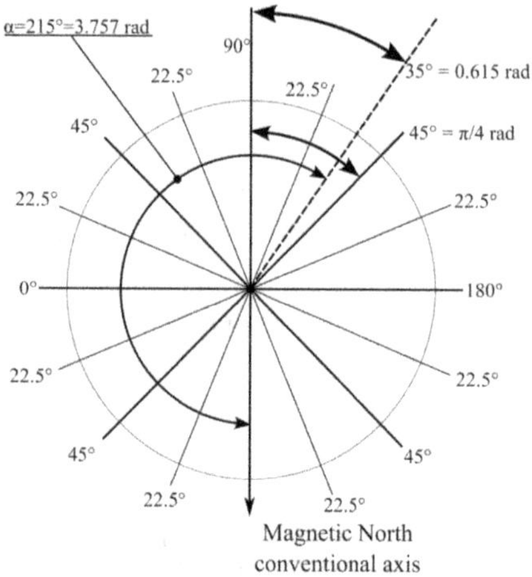

(a)

Fig. 19.4. (a) Goniometric reference frame sketching the angles of the LLI violation, according to the experiment of the magnetic coil, α and $\pi/4$, are shown the angles $\alpha = 215°$ and its reduction to first quadrant or rather $\alpha - 180° = 35°$.

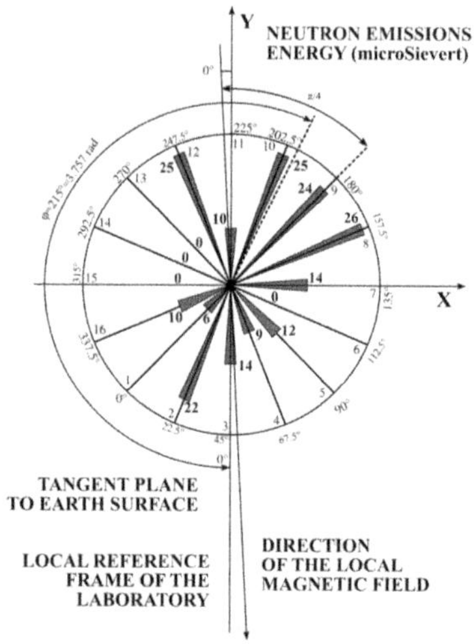

(b)

Fig. 19.4. (b) Polar diagram of the neutron emissions. The length of the bold lines starting from the center in the radial direction gives a visual description of the measured dose. This dose also includes the energy absorbed by the Teflon calorimeter, which reached the fusion temperature, $T_{\text{Teflon}} = 327°$C. The reference frame is according the convention of the angle law experiments, where the direction of local magnetic field is only conventional reference for the orientation of different experiments.

deformed space time DST theory and the connected angle law of Lorentz invariance LLI violation. As previously stated, the Teflon block in which the AISI 304 rod was inserted had the function of acting as a calorimeter. At the end of the experimental run the steel rod temperature, measured by an infrared thermometer Fluke 69 IR, was found to be 92°C, while the temperature of the Teflon block at the contact surface with the rod was 330°C, moreover as can be seen in Fig. 19.6 the Teflon appeared melted and somewhere carbonized. It was found that the energy consumed to bring the steel rod from the laboratory ambient temperature of 20°C to the final temperature of 92°C was equal to 6300 joules, and was absorbed by the ultrasonic generator from the power supply. Instead the energy absorbed by the calorimeter was estimated to be at minimum 62,500 joules.

This evidence forces to exclude an energy transfer due only to the applied ultrasound, but leads to assume a generation of it from DST nuclear reactions occurred in the steel bar during the sonication. These DST reactions were the origin of the observed neutrons, and could have produced a presumable dose of alpha rays, also responsible of the measured heat excess.

Again as we can see in Fig. 19.4, the PADC set up with the images obtained allowed us to get a 'stereoscopic' measurement of the energy of the produced neutrons as they outcome from the calorimeter. The neutron energy has been reported in a polar diagram, where we already pointed out as well evident the asymmetric and anisotropic distribution.

We want now to compare these results for the neutron energy emission, got in the Rome experiment, with the results of the angle law as obtained in the L'Aquila magnetic coil experiment reported in Chapter 13. Actually we superimposed the goniometric reference frame of Fig. 19.4(a) to the polar diagram of the neutron emissions obtaining the layout sketched in Fig. 19.4(b). From Fig. 19.4(b) now we test the match with the angle law. In order to verify this match we reported the whole situation in Fig. 19.5. At the center of Fig. 19.5 the development of the neutron emission, as measured by means of the CR39-ABs during the sonication experiment performed in Rome, is again reported. The steel bar emitting neutrons under ultrasound application was perpendicular to the plane tangent to the Earth surface. It is clear that the angle corresponding to the maximum of the asymmetric neutron emission is consistent with the angle of the

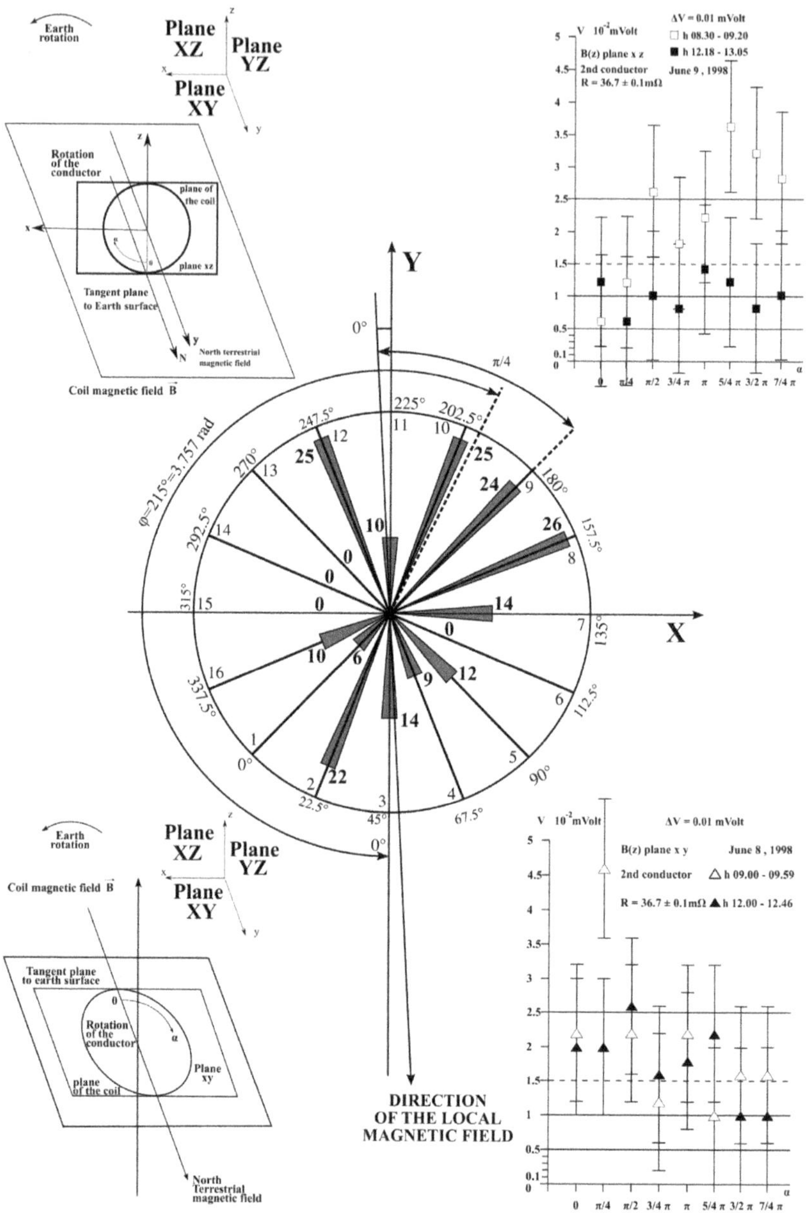

Fig. 19.5. Representation of the neutron emission in the Rome experiment at the center and of the coil experiment. The orientation of the neutron emission and of the voltage is normalized with respect to the direction of the local magnetic field.

maximum voltage measurement in the coil experiment. It is likewise evident that, to the maximum neutron emission, there corresponds a minimum emission exactly perpendicular to the maximum direction, as well as also for the Copper conductor in the coil. The two distinct experiments, the Rome one and the L'Aquila one, carried out in different conditions and some years later, involving two different classes of physical phenomena, electromagnetic and nuclear, both converge towards a clear geometrical violation of Lorentz invariance according the angle law, realizing what we defined at the beginning of Chapter 15 the 'repeatability in great'. Moreover, let us stress that the other Lorentz violation, corresponding to the angle $\pi/4$ on the (x, y)-plane, deserves to be taken into account. In fact, despite the possible incorrectness of its use, we can see that, according to the convention of the coil experiment, again the angle $\pi/4$ falls within the region of maximum neutron emission as we can see in Fig. 19.5. Namely, the maximum energy emission occurred at the arc including the two Lorentz breakdown angles. This clear agreement between a previous electromagnetic experiment and the neutron emission experiment leads us to conclude that the neutron emissions, and the DST reactions from which they originate, occur in conditions of Lorentz invariance violation, including its own asymmetry.

In Fig. 19.6, one of the calorimeters employed in the Rome experiments is shown. In all the experimental runs, the calorimeters showed a combustion asymmetrically extended in radial direction with respect to the vertical axis common to the steel bar and the calorimeter itself. Beside the energetic considerations already exposed, we want to examine in deep the correlation between the neutron emission and the calorimeter combustion. Figure 19.7 shows the neutron emission superimposed to the calorimeter. If we consider both the angles $5/4\pi$ and $\pi/4$, they show an arc which again falls within the region of both maximum neutron emission and maximum burnt area in the calorimeter. As a matter of fact, the correspondence between the Lorentz violation direction in the coil experiment and the maximum of neutron emission gives us the clue that Lorentz invariance and the angle law together rule the neutron emission and the consequent burnt of the calorimeter.

Let us recall that the energy spectrum of the emitted neutrons was measured along only one of the 16 considered directions, or rather the direction 16, see Fig. 19.3 at bottom left frame. It is clear that the

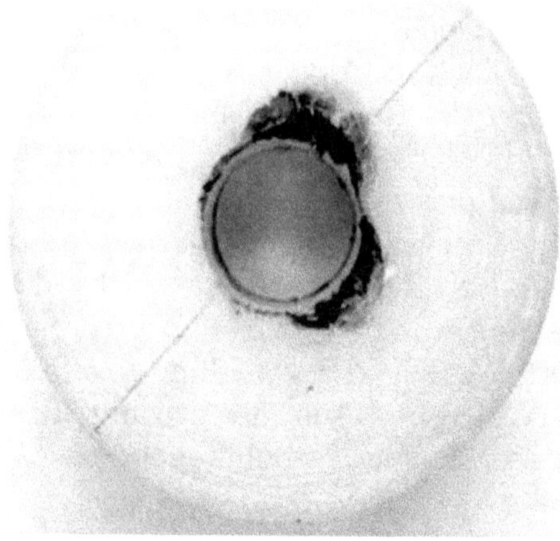

Fig. 19.6. Teflon cylinder which acted as calorimeter, with the burns corresponding to the preferential directions of the neutron emission.

energy release from neutrons in the calorimeter depends on their own energy, therefore, where combustion is present, it must be checked if there are neutrons, but a combustion does not always correspond to neutron presence. We think that the neutron energy was not sufficient to produce, in the crossed thickness of the calorimeter, an energy release adequate to yield a combustion. Furthermore, we are well aware that DST emissions include also α rays. The geometry of the experiment does not allow to perform α-ray measurements, but we cannot exclude that these also have been produced and concurred to the calorimeter combustion, thanks to their high linear energy transfer (LET). It must moreover be taken into account that in DST conditions a concomitant emission of α rays and neutrons was statistically verified, as explained in Chapter 7. As seen in Fig. 19.7, by using all the sixteen directions as reference, it was established that the correlation C1 between neutron emission and combustion is equal to 67%. Vice versa, the correlation C2 between combustion and neutron emission is 100%, the same percentage of the correlation C3 between no neutron emission and no combustion. Where the calorimeter is burnt, there is neutron emission, and where the calorimeter is not burnt there is no neutron emission. The difference between C1 and C2 is due to the energy spectrum of produced

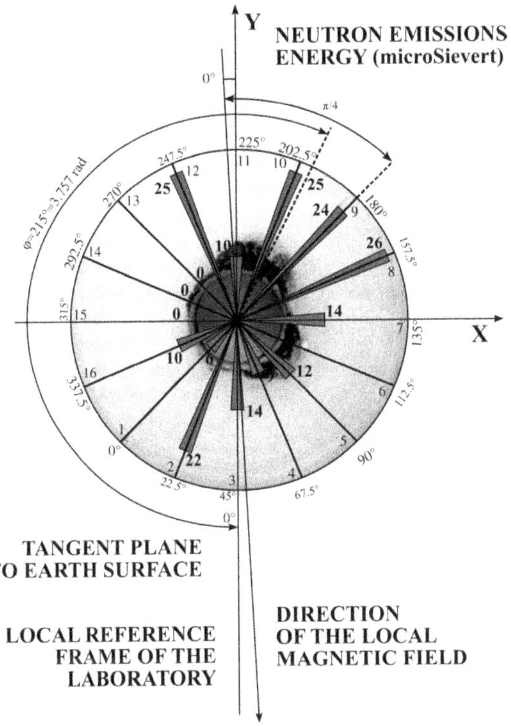

Fig. 19.7. Neutron emission superimposed to the calorimeter with the burn marks.

neutrons, namely if they are able or not to release energy in the calorimeter. We can therefore conclude that the calorimeter burning is a further proof of the asymmetry of the neutron emission which occurred according the angle law of Lorentz invariance violation [4].

The ultrasound pressure was symmetrical with respect to the axis of the cylindrical bar. Therefore, there is no apparent reason of connection between the asymmetrical neutron emissions and any macroscopic characteristics of asymmetry or inhomogeneity in the system constituted by the steel bar together with ultrasound generator. In conclusion, we tested the correspondence between the Lorentz violation asymmetry both in electromagnetic systems and the neutron emission asymmetry, as well as in alpha emissions. Thus, in both classes of phenomena electromagnetic and nuclear we get more stringent evidences of the fact that the DST conditions generated by

Lorentz violation are intimately connected with asymmetry, as a fundamental feature of Lorentz violation itself and that the angle law, which resumes this feature, has a general validity independent from the nature of the interaction ruling the phenomenon. In this sense, we assume that the angle law has a 'general' value as an independent 'natural law'. This evidence means that there is a fundamental asymmetry underlying all natural phenomena and any fundamental interaction. Now, we face the problem if the angle law has a value independent from the fact that it was tested in terrestrial experiments. Effectively if this angle law can connect phenomena at small scales, terrestrial, as well at large scales, cosmic, it can assume an 'universal' value too.

References

[1] A. Rosada *et al. Modern Phys. Lett. B*, 29, 14, 2015.
[2] A. Rosada and A. Petrucci, *J. Adv. Phys. Special Sect. Nuclear Reactions*, 5, 63–68, 2016.
[3] G. Cherubini *et al. Detection*, 1, 30, 2013.
[4] F. Cardone *et al. Eur. Phys. J. Plus*, 130, 55, 2015.

Chapter 20

Asymmetric Emissions and Cosmic Background Radiation Anisotropy

We examine the spatial anisotropies and asymmetries found in nuclear experiments on the violation of the local Lorentz invariance (LLI) exposed in Chapter 19 and their possible relationship with the spatial anomalies and anisotropies of the Cosmic Microwave Background Radiation (CMBR) in the Universe. We explore the possibility that there exists fundamental anisotropy and asymmetry which sustain both the CMBR and the space-energy level of the experiments on the LLI violation in the nuclear interacting systems [1].

Now, we test the correspondence between the anisotropies in the case of the nuclear emissions of neutrons, in condition of Deformed Space Time emission, and the CMBR preferred directions. Here we present the use of the same procedure applied to the electromagnetic case, which is examined in the next chapter.

Several types of spatial CMBR anisotropies have been identified by the Planck Surveyor mission, often confirming what was discovered by the precedent COBE and WMAP missions. The analysis and the results, based on the Planck mission, are the reported data from which are extracted the spatial directions of the anisotropies and then compared with those ones identified by analyzing the LLI nuclear experiments. This approach can provide enough evidence to support the hypothesis on the presence of anisotropies underlying both the CMBR and the experimental results as analyzed in Chapter 19, where there is evidence of a spatial asymmetry of the measured values of emitted neutrons, as well as an anisotropy of the values referred to the neutron measurements. In the CMBR, there is an anisotropy of the measured values that identifies an asymmetry of the directions and gives an indication of preferred directions.

The CMBR maps which have been used have been obtained from different types of spatial masking processes and component separation algorithms applied to the frequency maps of the released Planck in 2013 [2]. On those maps, statistical analyses have been carried out to identify the spatial anisotropies. Those analyses have lead to different results, with an associated statistical significance level with respect to the null hypothesis of isotropic Universe, for each identified anisotropy mainly dependent on the applied method. In order to summarize in one single result all the different results obtained for the same anisotropy, and considering that the obtained results are expressed in terms of spatial directions, have been used the spatial average of those directions having the highest statistical significance level. Then the so identified spatial anisotropies are the following ones:

- Alignment between the orientation of the Quadrupole and the Octopole components, determined by maximizing the angular momentum dispersion, hereafter denoted as QOA.
- Hemispherical Asymmetry, angular power spectrum asymmetry on the sky, hereafter denoted as HA.
- Power Asymmetry, dipolar power spectrum asymmetry on the sky, hereafter denoted as PA.
- Dipole Modulation, by power spectrum asymmetry expressed in terms of a multiplicative dipole modulation model, hereafter denoted as DM, and Doppler Boost, dipole modulation inducted by the Earth motion with respect to the CMBR rest frame, hereafter denoted as DB.
- Mirror Parity, mirror symmetry with respect to a plane, hereafter denoted by MP.
- Cold Spot, anomalous wide cold area in the CMBR sky, hereafter denoted as CS.

Table 20.1 shows the identified anisotropies and their galactic coordinates, longitude l and latitude b, that are uniquely associated with the corresponding directions.

Table 20.1. Directions in galactic coordinates of the identified CMBR anisotropies.

Anisotropy	QOA	HA	PA	DM	DB	MP	CS
			Galactic coordinates				
$l(°)$	234	237	193	227	264	264	209
$b(°)$	67	−20	−3	−15	48	−17	−57

The DST neutron emission experiment used as reference one was conducted in Rome the 19th of December 2012, 11.30 am. The spatial distribution of the neutron emissions energy is shown in Fig. 20.1, where it is evident both the asymmetry and anisotropy, together with the 'star-like' emission. This spatial distribution represents the experimental data to be compared with the CMBR anisotropies. In order to investigate the existence of identical preferred directions emerging from both the experiment here described and the CMBR sky map, it is necessary to describe the preferred directions found for the CMBR, see Table 20.1 in the same spatial reference system of the DST neutron experiment.

The spatial reference system, used to describe the directions of the neutron emissions, is an altazimuth system with the North represented by the magnetic North as in the case of the LLI coil experiment. The geographical North differs very slightly from the magnetic North. It was evaluated, at the site and the date of the experiment, that this difference is approximately 1°. Since the precision of the preferred directions of CMBR is equal to 1° of galactic latitude and longitude, we can consider the reference system of the experiments as an altazimuth system with the north given by the geographical North. We also consider fixed in the sky each CMBR privileged direction within 3 min duration of the experiment, the angular displacements in the sky of these directions during this time is negligible. Moreover, the applied statistical analysis was the Spearman test, and due to its sensitivity also to non linear correlation it is robust against the misalignment that affects the compared directions in a maximum time interval needed for the neutron measurements due to the Earth movements.

Therefore, the CMBR privileged directions can be transformed from galactic coordinates in altazimuth coordinates, true equator

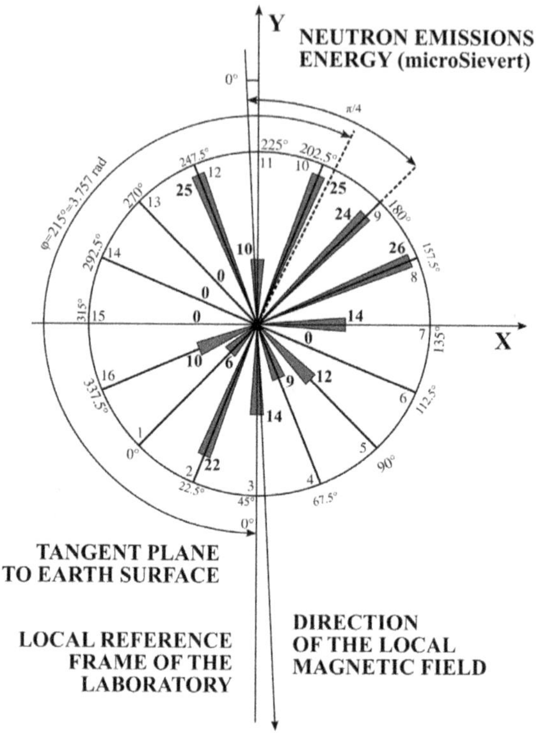

Fig. 20.1. Schematic view and measurements of the DST neutron experiment. The orientation of the neutron emission is referred to the direction of the local magnetic field.

and equinox of the date, into the geographical coordinates of the site, at the day and at the time of the experimental data recording.

By describing the directions of DST neutron emission and the directions of the CMBR anisotropies in the same spatial reference system, it is possible to quantitatively determine the alignment level by using the scalar product between vectors, where the vectors represent the considered directions. In this way, we can derive the values of the scalar products between all neutron emission directions and the spatial location of the CMBR anisotropies. It is clear that the direction of the CMBR anisotropies and anomalies is not contained in the horizontal plane where the direction of neutron emissions are contained. It has rather, an inclination with respect to this plane and the cosine values listed in Table 20.2 refer to angles in the space

Table 20.2. Values of the scalar products between DST neutron emission directions and CMBR directions of the different anisotropies, the angles in the top row are measured from the reference direction that coincides with the local magnetic field.

	0	$\pi/8$	$2\cdot\pi/8$	$3\cdot\pi/8$	$4\cdot\pi/8$	$5\cdot\pi/8$	$6\cdot\pi/8$	$7\cdot\pi/8$
AQO	−0.86	−0.72	−0.48	−0.16	0.18	0.50	0.73	0.86
HA	0.11	−0.20	−0.48	−0.69	−0.79	−0.77	−0.64	−0.41
PA	−0.44	−0.75	−0.94	−1.00	−0.90	−0.66	−0.32	0.06
DM	−0.04	−0.37	−0.63	−0.80	−0.85	−0.77	−0.57	−0.28
DB	−0.58	−0.48	−0.30	−0.08	0.16	0.37	0.52	0.60
MP	0.31	0.11	−0.11	−0.31	−0.46	−0.55	−0.55	−0.47
CS	0.48	0.11	−0.28	−0.62	−0.87	−0.99	−0.96	−0.78
	π	$9\cdot\pi/8$	$10\cdot\pi/8$	$11\cdot\pi/8$	$12\cdot\pi/8$	$13\cdot\pi/8$	$14\cdot\pi/8$	$15\cdot\pi/8$
AQO	0.86	0.72	0.48	0.16	−0.18	−0.50	−0.73	−0.86
HA	−0.11	0.20	0.48	0.69	0.79	0.77	0.64	0.41
PA	0.44	0.75	0.94	1.00	0.90	0.66	0.32	−0.06
DM	0.04	0.37	0.63	0.80	0.85	0.77	0.57	0.28
DB	0.58	0.48	0.30	0.08	−0.16	−0.37	−0.52	−0.60
MP	−0.31	−0.11	0.11	0.31	0.46	0.55	0.55	0.47
CS	−0.48	−0.11	0.28	0.62	0.87	0.99	0.96	0.78

above or below this plane, between this direction and each of the neutron emission directions.

Now, the hypothesis on the existence of an underlying anisotropy and asymmetry appearing simultaneously at the two experimental levels, which are the DST neutron experiment and the CMBR anisotropies, can be tested by analyzing the correlation of the measurements of the DST neutron experiment and the corresponding values of the scalar products.

This analysis provides correlation values, associated confidence levels, statistical p-values and statistical significance, with p-value ≤ 0.05.

The correlation analysis gives seven correlation values with an associated 95% confidence interval and a corresponding significance level. The Spearman rank correlation is used due to its robustness, see [3,4]. The correlation analysis is performed by taking into account two cases. First one using the direction and the orientation of the vectors identifying the directions of the CMBR anisotropies given by the scalar products with their sign. Second one only using the directions, given by the absolute values of the scalar products. A bootstrap method is used to determine the confidence intervals [5].

The results of the statistical analysis are reported in Tables 20.3 and 20.4. The former shows the correlation indices with their corresponding levels of significance, while the latter shows the confidence intervals only for the statistically significant correlation indices with a p-value less than 0.05.

The obtained results point to a negative correlation between the DST neutron experiment measurements and both the orientation and the direction of CMBR anisotropies Power Asymmetry (PA)

Table 20.3. Spearman correlation indices (r) and corresponding levels of significance (p_val) for the two cases. Columns 2 and 3: orientation and direction, scalar products with sign; columns 4 and 5: only direction, absolute values of scalar products.

	Orientation and direction		Direction	
CMBR anisotropy	r	p_val	r	p_val
QOA	−0.10	0.73	0.02	0.95
HA	−0.44	0.09	−0.42	0.11
PA	−0.53	0.04	0.19	0.48
DM	−0.49	0.05	−0.23	0.39
DB	−0.10	0.73	−0.01	0.97
MP	−0.21	0.44	−0.50	0.05
CS	−0.29	0.28	−0.62	0.01

Table 20.4. Confidence intervals, from r_min to r_max, for the statistically significant correlation indices based on Table 20.3.

	Confidence interval	
CMBR anisotropy	r_min	r_max
PA (orientation and direction)	−0.81	−0.04
DM (orientation and direction)	−0.82	0
MP (only direction)	−0.86	−0.03
CS (only direction)	−0.86	−0.12

and Dipole Modulation (DM). On the other hand, there is evidence of a negative correlation between the DST neutron experiment measurements and only the direction of the CMBR anisotropies Mirror Parity (MP) and Cold Spot (CS). This evidence is almost stronger for the Cold Spot anisotropy, for only direction, where both the significance of the statistical test and the correlation index are higher, the confidence interval is also further shifted with respect to the 0 value.

Therefore, the experimental measurements suggest that CMBR anisotropies were reproduced by nuclear emissions of DST neutrons. Moreover, as we will see in the next Chapter 21, the CMBR anisotropies were here reproduced with a stronger correlation with respect to the case of the LLI coil experiment. It is notable that a nuclear phenomena has a correlation with CMBR anisotropy stronger than an electromagnetic phenomena, but in both cases it is quite surprising that terrestrial scale experiments have so strong correlation with the properties of a cosmological scale phenomena.

Thus, we believe that these findings point to the possible existence of a common underlying asymmetry of the CMBR preferred directions and the Local Lorentz Invariance violation both in the electromagnetic and the nuclear phenomena as described and resumed by the angle law.

References

[1] F. Cardone *et al. J. Adv. Phys.*, 6(2), 303, 2017. DOI: 10.1166/jap.2017.1332.
[2] Planck Collaboration and E. Martinez Gonzaleza. *Astron. Astrophys.* DOI: 10.1051/0004-6361/201.321534.
[3] P. Wessa, Spearman Rank Correlation (v1.0.1) in Free Statistics Software (v1.1.23-r7), Office for Research Development and Education, 2012, http://www.wessa.net/rwasp_spearman.wasp/, last accessed 2017.
[4] AnswerMiner–Data Analysis made easy, https://www.answerminer.com/calculators/correlation-test/, last accessed 2017.
[5] R. I. Kabacoff, Nonparametric Bootstrapping, http://www.statmethods.net/advstats/bootstrapping.html, last accessed 2017.

Chapter 21

Lorentz Violation and Cosmic Background Radiation Anisotropy

We now investigate the spatial anisotropies and asymmetries emerging from the magnetic coil experiments on the violation of the local Lorentz invariance (LLI) and their potential relationship with the spatial anisotropies of the Cosmic Microwave Background Radiation (CMBR). The CMBR was first theorized by the physicist Gamow in 1948, in the context of the Big Bang cosmological theory, and it was discovered by the American physicists Penzias and Wilson in 1964. The CMBR is an electromagnetic radiation that fills our Universe as a black body spectrum at a temperature of 2.725°K [1], then the presence of underlying anisotropies, appearing at least in these two levels of space and energy, is suggested. Despite the fact that the CMBR can be assumed isotropic to roughly one part over 100,000, the presence of underlying anisotropies, appearing at least in these two levels of space and energy, is then suggested also in electromagnetic phenomena violating LLI, after the positive results on nuclear phenomena exposed in Chapter 20.

It is believed that the measured CMBR anisotropies correspond to temperature variations generated by quantistic fluctuations on small scales, such as that one in the Universe at the decoupling epoch, then increased in the space by the expansion of the Universe. This hypothesis allows a comparison of the CMBR anisotropies, at present Universe epoch, XXI century, with other anisotropies that might occur at smaller spatial scales. In this sense, we support the thesis that a fundamental anisotropy exists. This anisotropy would sustain both the CMBR and the space-energy level of the experiments on the LLI violation in the electromagnetic systems. Now, we wish to recap in brief a small history of those measurements which gave rise to the evidences of anisotropies and asymmetries in CMBR. The first measurements of the CMBR's spectrum and of the spatial anisotropies were done by the American satellite Cosmic Background Explorer, COBE, and the Russian Prognoz 9, performing the experiment RELIKT-1. The first maps of the CMBR with its spatial variation, obtained separately from COBE and RELIKT-1, were presented in 1992. Later, other maps of the CMBR with an increased angular resolution were released and many experiments have been performed to obtain high-resolution maps, here below we list some of the relevant ones. The terrestrial surface Degree Angular Scale Interferometer (DASI), an interferometric telescope that brought to the discover of the CMBR polarization in 2002. The Atacama Cosmology Telescope (ACT) and the South Pole Telescope (SPT), both radio telescopes with mirrors in Gregorian configuration. The space experiment Wilkinson Microwave Anisotropy Probe (WMAP), heir of the COBE satellite that ran until 2010. The Planck Surveyor, heir of the WMPA, with a wider electromagnetic spectrum, wider angular resolution and sensitivity [2].

The studies based on high precision data at a spatial resolution similar to the ones available from the Planck Surveyor mission, have identified several types of spatial CMBR anisotropies, confirming what was discovered by the COBE and WMAP missions. Again as for the DST neutron experiment the analysis and the results are based on the Planck mission, by using these data it is possible to extract the spatial directions of the anisotropies and compare them with those ones identified by analyzing the LLI violation in the coil experiments. This approach can provide evidence to support the hypothesis on the presence of anisotropies underlying both the CMBR and the experimental apparatus of the Local Lorentz Invariance coil experiment. As exposed in Chapter 13, in this experiment there is an asymmetry of the involved directions and an anisotropy of the measured values. In the CMBR, there is an anisotropy of the measured values that identifies an asymmetry of the directions and gives an indication of preferred directions.

CMBR maps have been obtained from different types of spatial masking processes and component separation algorithms applied to the frequency maps of the released Planck mission. On those maps, statistical analyses have been carried out to identify the spatial anisotropies. Those analyses have led to different results, with an associated statistical significance level with respect to the null hypothesis of isotropic Universe. It is clear that for each identified anisotropy it is mainly dependent on the applied method. In order to summarize in one single result all the different results obtained for the same anisotropy, and considering that results are expressed in terms of spatial directions, we derive the spatial average of those directions having the highest statistical significance level. The identified spatial anisotropies, employed to test the correspondence with LLI magnetic coil experiment, are the following ones:

- Alignment between the orientation of the Quadrupole and the Octopole components, determined by maximizing the angular momentum dispersion, hereafter denoted as QOA.
- Hemispherical Asymmetry, angular power spectrum asymmetry on the sky, hereafter denoted as HA.
- Power Asymmetry, dipolar power spectrum asymmetry on the sky, hereafter denoted as PA.
- Dipole Modulation by power spectrum asymmetry expressed in terms of a multiplicative dipole modulation model, hereafter denoted as DM, and Doppler Boost, dipole modulation inducted by the Earth motion with respect to the CMBR rest frame, hereafter denoted as DB.
- Mirror Parity, mirror symmetry with respect to a plane, hereafter denoted by MP.
- Cold Spot, anomalous wide cold area in the CMBR sky, hereafter denoted as CS.

Table 21.1 shows the identified anisotropies and their galactic coordinates, longitude l and latitude b that are uniquely associated with the corresponding directions.

Obviously this list is the same as in Chapter 20, but for the sake of completeness we have chosen to repeat it.

The LLI coil experiment used as reference was conducted in 1998, 8th and 9th of June. We remind that it consists in a Helmholtz coil with an axis orthogonal to a plane p containing a conductor with a capacitor. The conductor can change its spatial orientation in the plane p, and the whole experimental device can also turn in the space.

Table 21.1. Directions in galactic coordinates of the identified CMBR anisotropies, longitude l and latitude b are in degree (°).

Anisotropy	QOA	HA	PA	DM	DB	MP	CS
Galactic coordinates							
l (°)	234	237	193	227	264	264	209
b (°)	67	−20	−3	−15	48	−17	−57

A voltmeter connected in parallel with the capacitor can measure the voltage generated by a charge anisotropy along the conductor. A steady-state current circulating in the coil produces a constant magnetic field inside the coil, where the plane with the conductor and the capacitor is located. The orientation of the conductor in its plane is parametrized by an angle, ranging from 0 to 2π in steps of $\pi/4$. Then, the experimental data to be considered are the voltage measures. They are taken with the coil, the voltmeter, the conductor and the capacitor, in series with the conductor, at rest in the same inertial reference frame. In such an experimental condition, no Lorentz force, no voltage above the instrumental zero of the voltmeter, is expected on the device, according to the usual relativistic electrodynamics. However, as we widely illustrated, some measurements are outside the zero signal band. Figure 21.1 shows the layout of the experimental apparatus, together with the obtained measurements. The resulting anisotropies of the measurements as well as the related asymmetries with respect to the spatial directions are quite evident, as shown in Fig. 21.1(d), 21.1(e).

For a geometric and spatial description of the experimental apparatus, it is necessary to define a Cartesian coordinate system, which we recall. The y axis coincides with the projection on the horizontal plane of the Earth's local magnetic field direction, with positive direction, called North, towards the local magnetic North Pole, the horizontal plane is tangent to the Earth's surface in the position of the experimental apparatus. The z axis, perpendicular to the y axis and to the horizontal plane, has positive direction towards the zenith. The x axis, perpendicular to the y axis and to the z axis, with positive direction, is called East, to see a counterclockwise rotation from the y axis to the z axis. The center of the geometrical coordinate system coincides with the center of the coil.

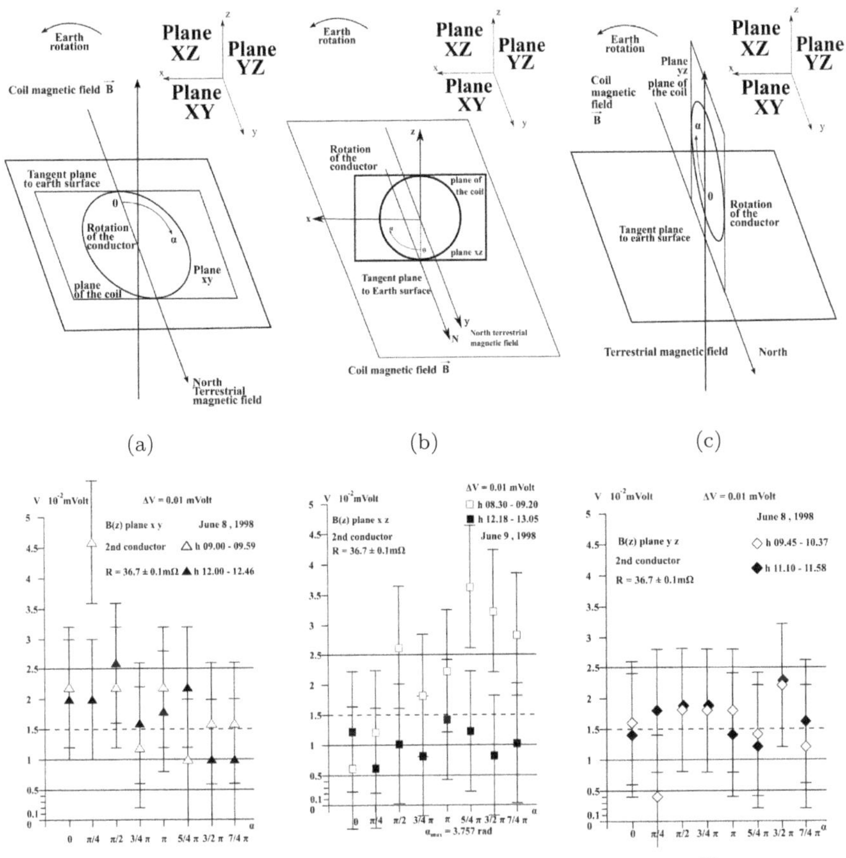

Fig. 21.1. Schematic view and measurements of the LLI coil experiment; the LLI violation and its asymmetry is well evident in the plots (d) and (e) which refer respectively to the xy and xz planes. (a) Schematic view of the orientation of the apparatus and the related magnetic fields in the xy plane. (b) Schematic view of the orientation of the apparatus and the related magnetic fields in the xz plane. (c) Schematic view of the orientation of the apparatus and the related magnetic fields in the yz plane. (d) Results of the measurements with the apparatus in the xy plane. $\Delta\text{V} = 0.01\,\text{mV}$ is the error on the voltage V across the capacitor. (e) Results of the measurements with the apparatus in the xz plane. $\Delta\text{V} = 0.01\,\text{mV}$ is the error on the voltage V across the capacitor and (f) Results of the measurements with the apparatus in the yz plane. $\Delta\text{V} = 0.01\,\text{mV}$ is the error on the voltage V across the capacitor.

The conductor step-rotation, not continuous, as said is done in steps of $\pi/4$ over the whole plane angle and there are three different positions of the plane p of the coil, according to the canonical planes of the orthogonal Cartesian system: plane coincident with the plane defined by the xy axes, plane coincident with the plane defined by the xz axes and plane coincident with the plane defined by the yz axes. In the case of the plane defined by the xy axes, the rotation of the conductor is clockwise when viewed in the half-space above the Earth's surface, and starts, zero angle, with the axis of the conductor along the negative y axis. In the case of the plane coincident with the plane defined by the xz axes, the rotation of the conductor is clockwise when viewed from the North, and starts with the axis of the conductor directed as the negative z axis. In the case of the plane coincident with the plane defined by the yz axes, the rotation of the conductor is clockwise when viewed from East, and starts with the axis of the conductor directed as the negative z axis.

In order to investigate the existence of identical preferred directions emerging from both the experiment here described and the CMBR sky map, it is necessary to describe the preferred directions found for the CMBR, see Table 21.1, in the same spatial reference system of the LLI coil experiment.

The spatial reference system used to describe the experimental apparatus is an altazimuth system, in which the North is the magnetic North. An altazimuth system for astronomical use has the geographic North rather than the magnetic North, because they slightly differ. This difference was evaluated using an application of the NASA [5] so obtaining the value of the magnetic declination, the angle in the horizontal plane between the magnetic North and true North, as well for the space and time coordinates of the site of the experiment. This value is about $+1.4°$, where the plus sign indicates east deviation. It is worth noting that the accuracy in the positions of the conductor during its change of direction is definitely higher than $1°$ and the positioning accuracy of the planes of the coil is definitely around or above $1°$. The precision of the preferred directions of CMBR is equal to $1°$ of galactic latitude and longitude therefore we consider fixed in the sky every CMBR's privileged direction during each data recording which lasted about 50 minutes. Thus, the reference system used in the experiment can be considered a geographic north altazimuth reference system.

Table 21.2. Values of the scalar products between LLI coil experiment conductor's direction and CMBR's anisotropies directions, obtained for the experiment of the 9th of June 1998, 8:30 am–9:20 am, with the coil in the xz plane.

	0	$\pi/4$	$2\cdot\pi/4$	$3\cdot\pi/4$	$4\cdot\pi/4$	$5\cdot\pi/4$	$6\cdot\pi/4$	$7\cdot\pi/4$
QOA	0.44	0.61	0.42	−0.01	−0.44	−0.61	−0.42	0.01
HA	0.19	0.74	0.85	0.47	−0.19	−0.74	−0.85	−0.47
PA	−0.36	0.41	0.93	0.91	0.36	−0.41	−0.93	−0.91
DM	0.07	0.71	0.93	0.61	−0.07	−0.71	−0.93	−0.61
DB	0.73	0.85	0.47	−0.18	−0.73	−0.85	−0.47	0.18
MP	0.56	0.82	0.61	0.03	−0.56	−0.82	−0.61	−0.03
CS	−0.36	0.10	0.49	0.60	0.36	−0.10	−0.49	−0.60

Table 21.2 reports the values of the scalar products between LLI coil experiment conductor's direction and CMBR's anisotropies directions, obtained for the experiment of the 9th of June 1998, 8:30 am–9:20 am, with the coil in the xz plane. It is possible to consider fixed the CMBR's privileged directions during the time interval for the measurements, 6 in about 50 minutes, because the angular displacements in the sky of these directions during this time is less than 4 times the angular resolution with which the measurements are taken. Moreover, the applied statistical analysis is robust against the misalignment that affects the compared directions and intervenes in a maximum time interval such as the one needed for the measurements. Therefore, the CMBR's privileged directions were transformed from galactic coordinates in altazimuth coordinates, or rather true equator and equinox of the date for the geographical coordinates of the site of the experiment, in the days of the experiment, and at the intermediate time of each data taking. Similarly, the various conductor's directions were expressed in the same altazimuth coordinates, in the x, y, z system.

By describing the directions of the conductor of the LLI coil experiment and the directions of the CMBR's anisotropies in the same spatial reference system, it is possible to quantitatively determine the alignment level by using the scalar product between vectors, where the vectors represent the considered directions. In this way, it is possible to derive the values of the scalar products between all possible pairs of conductor's position and the spatial location of the CMBR's anisotropy. All these values can be summarized in six tables.

Table 21.3. For all the CMBR's anisotropies, Spearman's correlation indices (r) and corresponding levels of significance (p_val), in the cases: orientation and direction, scalar products with sign, columns 2 and 3; only directions, absolute values of the scalar products, 'columns 4 and 5.

	Orientation and direction		Direction	
CMBR anisotropy	r	p_val	r	p_val
QOA	−0.346	0.016	0.020	0.892
HA	−0.190	0.196	0.272	0.061
PA	−0.128	0.386	−0.004	0.980
DM	−0.183	0.212	0.243	0.097
DB	−0.412	0.004	−0.120	0.417
MP	−0.240	0.100	0.095	0.521
CS	0.079	0.593	0.307	0.034

Table 21.3 reports the results related to the experiment performed the 9th of June 1998, from 8:30 am to 9:20 am.

Now, the hypothesis on the existence of an underlying anisotropy appearing simultaneously at the two experimental levels, LLI coil experiment and the CMBR anisotropies, can be tested by analyzing the correlation of the time series from the LLI coil experiment and the corresponding values of the scalar products. This analysis can provide correlation values, associated confidence levels, the statistical p-values, and the statistical significance, here p-value ≤ 0.05. Despite the fact that for brevity we show here the example of the measurements performed on the 9th of June 1998 all the series of measurements during the repeated experiments, in the years 1998–2000, were taken in exam by the statistical methods described and compared. The series which have been compared are those ones obtained by merging all the 6 sets of 8 experimental measures, thus 48 values. Each of the 7 series of scalar products, where 7 is the number of the different investigated anisotropies, has 48 elements which are the scalar products. The series are built by concatenating the 8 subseries according to the temporal order. Therefore, the correlation analysis leads to 7 correlation values with an associated 95% confidence interval and a corresponding significance level. The Spearman rank correlation was used because the samples are not Gaussian distributed and focuses on the relative ordering. As in the DST neutron experiment, the correlation analysis was performed by taking into account both the direction and the orientation of the vectors identifying the directions of the CMBR's anisotropies, the scalar products with their sign, and only the directions, or rather the absolute values of the scalar products. The bootstrap method was used to determine the confidence intervals.

The results of the statistical analysis are reported in Tables 21.3 and 21.4. The former shows the correlation indices with their corresponding levels of significance, while the latter shows the confidence intervals only for the statistically significant correlation indices with p value less than 0.05. The obtained results point at a negative correlation between the orientation and the direction of the LLI coil experiment measurements and the CMBR's anisotropies QOA, Quadrupole and Octupole alignments, and DB, Doppler boost. While, there is evidence of a correlation between the LLI coil experiment measurements and the direction of the CMBR's anisotropy CS Cold Spot. This evidence is slightly stronger for the DB anisotropy, in orientation and direction, where both the significance of the statistical test and the correlation index are higher, the confidence interval is also further shifted with respect to the 0 value.

Therefore, the experimental measurements suggest that CMBR anisotropies have been reproduced, and in particular the alignment between the orientation of the the Cold Spot, and also the Quadrupole and the Octopole components, the dipole modulation inducted by the Earth motion with respect to the CMBR rest frame Doppler Boost. We believe that these findings point at the existence of a common underlying asymmetry of the CMBR's preferred directions and LLI violation [3]. It is a relevant result the fact that the

Table 21.4. Confidence intervals from r_min to r_max for the statistically significant correlation indices based on data reported in Table 21.3.

CMBR anisotropy	Confidence interval	
	r_min	r_max
QOA (orientation and direction)	−0.646	−0.041
DB (orientation and direction)	−0.688	−0.135
CS (only direction)	0.035	0.586

Cold Spot direction is the CMBR preferred direction which shows correlation with both electromagnetic experiment and nuclear one, or rather with the LLI violating magnetic coil experiment and the LLI violating DST neutrons experiment [4]. Whereas the DST neutron experiment shows a stronger correlation with the Cold Spot direction, the magnetic coil seems to have a weaker one. For this reason we consider if there are any spurious effects in this experiment, since we want to clear the ground from any doubt which could distort or dim the possibility that there are actually evidences which support whether the LLI violation ruled by general angle law holds for any fundamental interaction, or whether there exists a fundamental asymmetry underlying both terrestrial and cosmic phenomena, all connected by an angle law having universal value.

In simple words we consider if there exists an angle law valid for each interaction or exist an universal angle law regardless any interactions.

In other words there are two possibilities: in the first one the fundamental physical laws, or rather the interactions, follow the angle law and in this case it must be implemented in each interaction in different mathematical proper way; in the second one the angle law has an universal value and therefore holds in the same way for all the interactions, as a consequence its mathematical implementation occurs in the same way for all interactions, this means that the angle law is a true fundamental physical law, overwhelming all the others ones.

We recall that in experiment performed in 1999–2000 with the same experimental device of the LLI coil experiment, 7 measurements were taken with two different values of the magnetic field inside the coil, and almost similar values of voltage were obtained thus excluding that the effect has been caused by Lorentz forces on the conductor. Moreover reversing the sign of the current in the circuit of the coil, no change in the sign on measured voltages occurred. External effects on the experimental apparatus might have affected the measurements, especially with regard to changes in the Earth's magnetic field and the secondary cosmic rays. Among those effects, there might be a diurnal variation of the magnetic declination, which typically reaches a maximum amplitude of less than 10 arc minutes, well below the instrumental precision, and however occurring at time scales larger than the experimental data taking. The diurnal variation, which is subject to the magnetic inclination, has also a mean amplitude well below the instrumental precision of the measurements. The anomaly of the diurnal variation of the magnitude of the magnetic field is very small, in the order of some tens of nanoTesla, maximum few hundreds, and occurs at time scales greater than those ones needed for data taking [6]. Effects of the magnetic induction within the experimental apparatus due to variations, at small time scales, of the local terrestrial magnetic field can therefore be excluded. We can also consider an interaction of the secondary cosmic rays with the experimental apparatus, in particular by positive muons of cosmic origin, of which there should be an evidence of an east–west anisotropy [7] generating a surplus of electrons and then an additionally charge across the capacitor, working as the revealing surface. But the scarcity of muons and especially the limited revealing surface, as well as the low integration time, corresponding to the data taking time, lead us to assume that this effect is completely negligible.

References

[1] D. J. Fixsen, *Astrophys. J.*, 707, 916, 2009.

[2] Planck Collaboration and E. Martinez-Gonzaleza. Planckn results XXIII *Astronomy and Astrophysics*, DOI: 10.1051/0004-6361/201321534.

[3] F. Cardone *et al. J. Adv. Phys.*, 4(2), 155–159, 2015.

[4] F. Cardone Lorentz Violation and CMRB Anisotropy, *Proceedings of the Xth Symposium 'Vigier'*, Ancona, Italy, July 2016. Unified Field mechanics: Formulation end empirical tests, section V. Astrophysics and Cosmology p. 369–374 Ed. World Scientific, 2018. ISBN: 978-981-3232-03-y/04-4/05-1.

[5] http://ccmc.gsfc.nasa.gov/modelweb/models/igrf_vitmo.php, last accessed 2017.

[6] V. Courtillot, J. L. L. Mouel, *Annu. Rev. Earth Planet. Sci.*, 16, 1998.

[7] 11. P. F. F. Grieder, *Cosmic Rays at Earth—Researcher's Reference Manual and Data Book*, Ed. Elsevier, 2001.

Chapter 22

Fundamental Asymmetry and Copernican Principle

In this chapter, we recall all the results and evidences giving rise and basis to consider the angle law of Lorentz violation whether a general law, common to both electromagnetic and nuclear phenomena, or whether an universal law, common to terrestrial and cosmic phenomena. Thus, the conclusion we can put forward is that a fundamental asymmetry underlies all the physical phenomena at a geometric dimensional level higher than four physical dimensions.

After about one century of validity tests of the local Lorentz invariance, LLI, we can summarize the results by the graph showed in Fig. 22.1 presenting its upper limits of violation. Several types of experiments are reported: Michelson–Morley type experiment, isotropy of the speed of light, tests of time dilation, nuclear magnetic resonance (NMR) test of the nuclear levels. Two intervals of LLI violation upper-bound are evident between 10^{-10}–10^{-2} and between 10^{-22}–10^{-14}. According to the deformed space time, DST theory, these two intervals are energy range controlled through the interaction working in the test apparatus at fundamental level of the phenomenon exploited. In any case the results reported in Fig. 22.1 refer to a kinetic Lorentz violation so introducing a drift velocity classical premise of a preferred reference frame, at least within the framework of the usual Einsteinian mechanics.

In the magnetic coil experiments, schematically reported in Figs. 22.3 and 22.4 and described in details in Chapter 13, the voltage inducted in the conductor by the static magnetic field is compatible

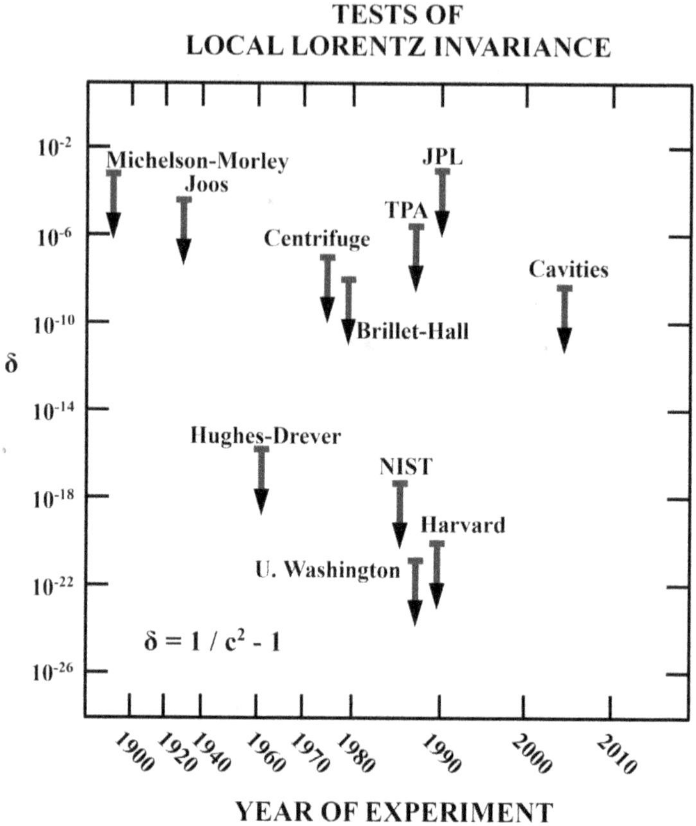

Fig. 22.1. Summary of tests for Local Lorentz Invariance.

Fig. 22.2. Anisotropy of the CMBR showing in galactic coordinates the presence of anomalies in the CMBR temperature.

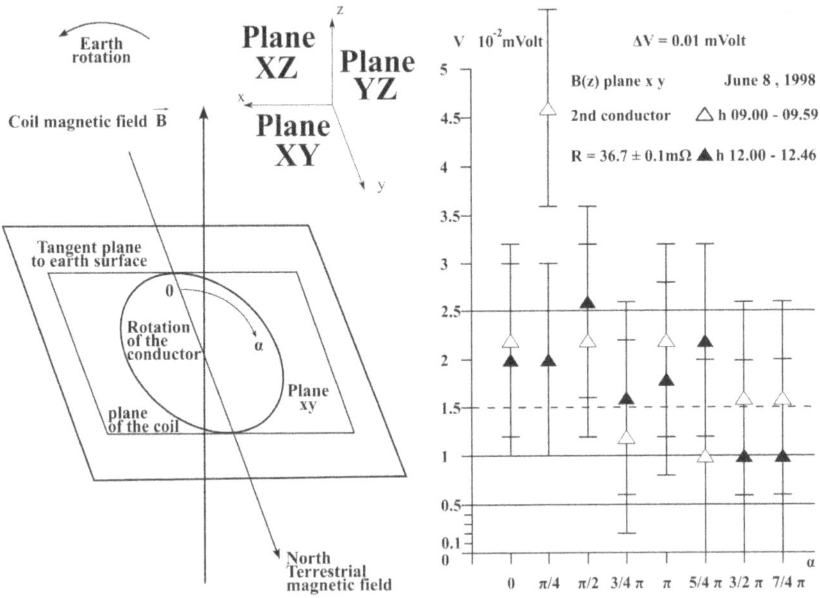

Fig. 22.3. LLI violation measurement with the apparatus in the xy plane.

with the drift velocity of 0.06 m/s which means compatible with zero considering the velocity of Earth and that of the Galaxy. Therefore, there is no kinetic violation of LLI and there is in no case any preferred reference frame. However, the LLI violation that does exist because there is an induced voltage, we strongly stressed to be only a directional violation and therefore indicating the existence of an asymmetry and anisotropy at least in the space.

The graphs in Figs. 22.3 and 22.4 show the existence of two directions along which LLI is violated.

xy plane – $\alpha = \pi/4$ with respect to the direction of the north magnetic field of Earth

xz plane – $\alpha = 5\pi/4$ with respect to the direction of the north magnetic field of Earth

yz plane – no violation

where the angle α refers to the direction of the conductor with respect to the earth magnetic field.

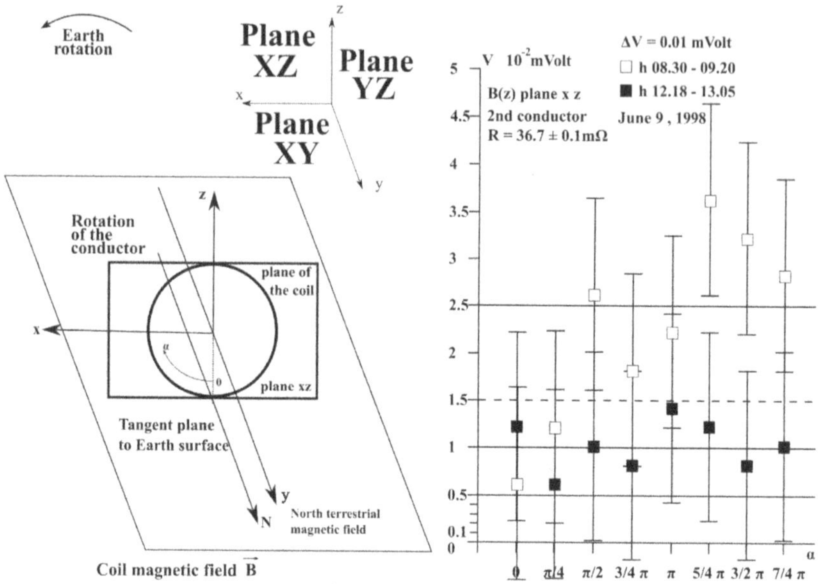

Fig. 22.4. LLI violation measurement with the apparatus in the xz plane.

Since LLI violation has an asymmetry, which implies that there are some privileged directions in space, in previous chapter we wanted to compare these directions with those existing for the cosmic microwave background radiation CMBR. In order to do this, we transformed the CMBR privileged directions, see Fig. 22.2, from galactic coordinates into the altazimuth coordinates of the geographical position of the site where the coil experiment was performed (L'Aquila-Italy), in the days when it was carried out, at the beginning of June 1998, and at the intermediate times of each data taking run.

The directions of the conductor in the coil experiment are ascribed to the same altazimuth coordinates. Once in the same spatial reference system, we compared the alignment level of the vectors, CMBR anomalies/anisotropies and LLI violation directions, by taking their scalar product.

In the previous chapter, we reported the tested hypothesis of an underlying anisotropy and asymmetry appearing simultaneously at the two experimental levels, LLI experiment and the CMBR anisotropies. The result of the correlation of the time series from the LLI experiment and the corresponding values of the scalar products are summarized in Tables 22.1 and 22.2. These tables show the Spearman correlation indices (r) and the corresponding levels of significance (p_val) for the two cases: orientation and direction of the scalar product, and only direction, the absolute value of the scalar products.

The obtained results point towards a correlation, of negative sign, between the direction and the orientation of the LLI experiment measurements and the CMBR's anisotropies for the QOA, Quadrupole and Octupole alignments and DB, Doppler boost, anomalies.

On the other hand, there is evidence of a positive correlation between the LLI experiment measurements and the direction of the CMBR anisotropy CS, Cold Spot. We consider this second correlation as fundamental from a physical perspective, since the Cold Spot is not accountable so far in term of any model. This correlation points towards a possible existence of a common underlying asymmetry of the CMBR preferred directions and the directions of LLI violation.

Table 22.1. Spearman correlation indices (r) and corresponding levels of significance (p_val) for the two cases. Columns 2 and 3: orientation and direction, scalar products with sign; columns 4 and 5: only direction, absolute values of scalar products.

	Orientation and direction		Direction	
CMBR anisotropy	r	p_val	r	p_val
QOA	−0.346	0.016	0.020	0.892
HA	−0.190	0.196	0.272	0.061
PA	−0.128	0.386	−0.004	0.980
DM	−0.183	0.212	0.243	0.097
DB	−0.412	0.004	−0.120	0.417
MP	−0.240	0.100	0.095	0.521
CS	0.079	0.593	0.307	0.034

Table 22.2. Confidence intervals, from r_min to r_max, for the statistically significant correlation indices based on Table 20.3.

	Confidence interval	
CMBR anisotropy	r_min	r_max
QOA (orientation and direction)	−0.646	−0.041
DB (orientation and direction)	−0.688	−0.135
CS (only direction)	0.035	0.586

As accounted in Chapter 19 the evidences of the possible violation of LLI under the angle law extended also to the nuclear interaction. In particular, it is possible to generate deformed space time conditions in matter by locally concentrating a suitable amount of energy in space and time, i.e. reaching a suitable energy density in space, Pascal or J/m^3 and in time, Watt. As described in Chapter 19 the DST nuclear reaction produced by energy density gave rise to the DST emitted neutron pattern shown in Fig. 22.5. In the same figure it is also reported the coincidence of the direction of LLI violation and the maximum of the anisotropic emission of neutrons.

xy plane – $\alpha = \pi/4$ with respect to the direction of the north magnetic field of Earth

xz plane – $\alpha = 5\pi/4$ with respect to the direction of the north magnetic field of Earth

The same type of investigation about the possible correlation between the directions of LLI violation and the directions of the anisotropies of the CMBR has also been performed between the direction of DST neutron emission and the CMBR. As already stated in both Chapters 20 and 21, in order to investigate the existence of identical preferred directions emerging from both phenomena it is necessary to describe the preferred directions found for the CMBR in the same spatial reference system of the DST neutron experiment, which for its part is obviously the same as the magnetic coil experiment. So, all the comparisons among the three physical situations are quite well acquired and give us a complete verification of the angle law validity.

Then from Chapter 20, we recall that the spatial reference system, used to describe the directions of the neutron emissions, is an altazimuth system with the North represented by the magnetic North. The geographical North differs very slightly from the magnetic North. At the site and the date of the experiment, this evaluated difference is approximately $1°$. Since the precision of the preferred directions of CMBR is equal to $1°$ of galactic latitude and longitude, we can consider the reference system of the experiments as an altazimuth system with the North given by the geographical North. We also considered fixed in the sky each CMBR privileged direction within 3 min duration of the experiment, as well as the angular displacements in the sky of these directions during this time is negligible. So, the CMBR privileged directions can be transformed from galactic coordinates in

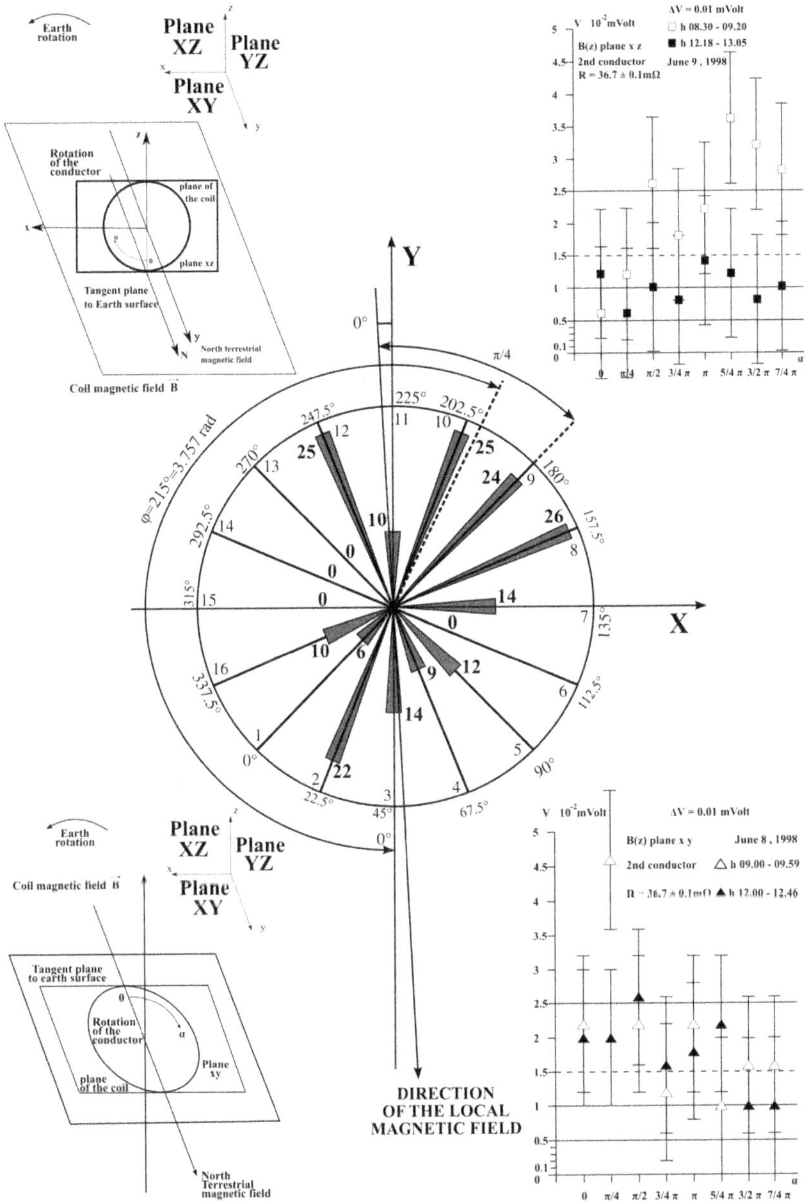

Fig. 22.5. DST neutrons. The emitted neutron pattern at the centre of the figure is highly anisotropic and antisymmetric. The four drawings of diagrams on the four corners, represent the violation of LLI in the electromagnetic experiment of Figs. 22.3 and 22.4.

altazimuth coordinates for the geographical coordinates of the site, at the day and at the time of the experiment data recording. Once the same reference system is applied to the directions of neutron emission and of the anomalies of CMBR, it is possible to calculate the scalar product of the angles between the direction of anomalies and the 16 directions of neutron emission. A Spearman correlation test was carried out between the values of the scalar product and those of the micro Sievert of emitted neutron. In Table 22.3, the indices (r) of the Spearman test and the corresponding levels of significance are reported for the two cases, orientation and direction of the scalar product, and only direction, the absolute value of the scalar products.

Beside Power Asymmetry (PA) and Dipole Modulation (DM), there is evidence of a correlation, negative, between the DST-Neutron experiment measurements and the direction of the CMBR anisotropies MP, Mirror Parity, and CS the Cold Spot.

This evidence is slightly stronger for the Cold Spot (CS) anisotropy only for direction, where both the significance of the statistical tests and the correlation index are higher, moreover the confidence interval is further shifted with respect to the zero value, see Table 22.3.

Therefore, these evidences suggest, as in the case of the LLI violation in magnetic coil experiment, that there exists a common alignment between the directions of the CMBR anisotropies and those of the nuclear emissions of DST neutrons.

Moreover, it appears that the correlation between the CMBR anisotropies and those of the DST neutron emission is stronger than the correlation with the magnetic coil experiments. For the sake of

Table 22.3. Confidence intervals from r_min and r_max for the statistically significant correlation indices of the Spearman correlation for DST neutron experiment as in Fig. 22.5.

	Confidence interval	
CMBR anisotropy	r_min	r_max
PA (orientation and direction)	−0.81	−0.04
DM (orientation and direction)	−0.82	0
MP (only orientation)	−0.86	−0.03
CS (only orientation)	−0.86	−0.12

completeness, the Spearman coefficient was used since we had to deal with non-Gaussian data having different variance, as in DST neutrons and Helmholtz magnetic coil experiments. In fact, this index is well suitable because it does not depend on adjustable parameters. Moreover, it does not search for the linearity of the response among the variables under test but it looks only for the common variation since it is more general not to consider the measured variables as possessing a linear interdependence. Anyway, also the Pearson index was used and it gave comparable results. For this reason, it is possible to rely on the robustness of the statistical tests performed which bring us to the conclusion that it appears to exist a strong correlation between terrestrial scale experiments and the properties of cosmological scale phenomena

Therefore, in the light of all these results and evidences of a common asymmetry among terrestrial scale experiments and cosmological scale phenomena we can do the following summary:

- Asymmetry in electromagnetic LLI violation giving the angle law.
- Anisotropy in DST neutron emission following the angle law and giving its asymmetry according LLI violation.
- Anisotropy in CMBR whose Cold Spot direction matches the angle law in the previous two LLI violations either electromagnetic or nuclear.

This 'triangular' coincidence is in the methodological frame of a double cross check as already stated in the whole experimental and phenomenological exams of the physical occurrences showing 'directional violation' of the LLI in terrestrial tests. Now, we can draw the consequences of all these occurrences besides terrestrial scale even to cosmic scale too.

Despite the well grounded 'Copernican Principle', stating that Earth is not a privileged observation point in the Universe, the evidences presented so far seem to point towards the need of a deep revision. Earth is neither a privileged point nor a non-privileged point of observation in the Universe, rather, it is a point of observation 'conditioned by natural laws themselves', which appear to possess a common asymmetry. In this sense, the mathematical description of Nature needs no longer a spontaneous symmetry breaking, it does not need any symmetry at all despite its usefulness in solving the mathematics of physics, like Noether theorem since 1918, as we already stigmatized in Chapter 16. All it is needed is Lorentz violation and to understand the underlying asymmetry in the natural laws of interactions. In a more cautious way we can simply state that the laws of Nature are already endowed with their own asymmetry, belonging to interactions through the energy they exchange [1].

Now at the end of this chapter, we wish to do a backward step since we are well aware of the consequences that such evidences can cause on a 'so-called' isotropy principle which states that do not exist privileged directions in the Universe as well does not exist any privileged reference frame. This isotropy principle is at the very basis of any theory of an 'Isotropic Universe' which includes explicitly a space endowed with the properties of isotropy 'the space is the same in all directions', thus preserving the angular momentum conservation, and homogeneity 'the space is measured in the same way in all directions', thus preserving the linear momentum. Moreover, this theory includes implicitly a time homogeneous, with the space, thus preserving the time momentum, or the energy. But the isotropy principle has another feature as consequence of the non-existence of any privileged direction, the fact that do not exist any privileged vectorial fields having any geometrical or physical meaning. Therefore, ever in an Isotropic Universe theory the gradient of any scalar field is null and from the property of homogeneity the scalar fields are simply constant on any space section of this universe and they are functions only of a 'so-called' cosmic time, see for example [2].

Now, the evidences and cosmic consequences of the angle law of the LLI violation are besides the DST theory which ties energy with space and time in a geometrical formulation of fundamental natural laws: the interactions. These evidences do not pave the way but in a certain sense open a way to use vector fields no longer as only mathematical tools but also as a physical possibility grounded on the angle law within a cosmic privileged direction: the direction of the cold spot in the CMBR.

References

[1] F. Cardone *et al. J. Adv. Phys.*, 7(1), 17–25, 2018. DOI: 10.1166/jap.2018.1389.

[2] S. Benenti, *Mathematical Foundations and Numerical Analysis of the Dynamics of an Isotropic Universe*, Ed. Springer, 2024.

Chapter 23

Conclusions and Consequences

23.1 Geometric Lorentz Violation

We have seen how the geometric violation of Lorentz invariance leads to a five-dimensional framework, in contrast to the kinematic violation of Galilean invariance, which results in a four dimensional structure. From the necessity to recover a new scenario within a physical context that incorporates both new transformations beyond those of Galilean relativity and a new symmetry, two historical choices emerge.

Lorentz's choice: each interaction is associated with its own coordinate transformations and can therefore possess its own metric.

Einstein's choice: all interactions share the same coordinate transformations and can thus be described by the same metric; this is the Einsteinian special relativity, which leads to the Einsteinian mechanics, excluding the gravitational interaction, which has got its own distinct metric.

In this volume, we have followed Lorentz's choice in order to arrive at a five-dimensional formulation of the spacetime deformation theory, the DST Theory.

23.2 Violations Recovered

From Lorentz's choice and the resulting DST Theory follows the necessity to address the recovery of Galilei and Lorentz violations through an extended, physically measurable coordinate as an additional dimension; both the choice of time as the fourth dimension and the choice of energy as the fifth dimension are justified a posteriori on experimental grounds. In the case of time, we have taken into account the well known fact that time measurements depend both on the state of motion and on the position within a gravitational field. In the case of energy, we started from the observation that its distribution in ellipsoidal form can be reduced to a spherical one, as confirmed by the torsional antenna, where the polar diagram of the energy distribution can be reduced to a circle. All of this provides an experimental recovery of a symmetry by means of the additional physical dimension introduced to account for the violation, that is energy. However, at this stage, this framework does not yet lead to new, equivalent Lorentz transformations; thus, we have limited our analysis to the experimental observation, which allows for the recovery of a symmetry in the energy distribution even in the presence of a deformed energy space time structure.

We wish to emphasize once again the distinction between the four dimensional kinematic Einsteinian framework and the five dimensional geometric Lorentzian framework. In the Einsteinian case, it is possible to write coordinate transformations within a four dimensional setting. In the Lorentzian case, within a deformed four dimensional space, one can still write coordinate transformations using a metric that depends on energy; while, in the five dimensional framework where energy is the fifth dimension, it remains an open question how to formulate coordinate transformations from which the experimentally observed symmetry might emerge.

23.3 Energy Densities

We saw in Chapter 12 that for the four fundamental interactions known there exists the spatial critical energy density calculated by the thresholds energies and the critical volume, useful for the DST reactions, which has been taken as unique critical volume. Now, we consider how such critical energy densities have to be compared with the spatial energy density in the Universe, which is estimated by cosmological measures. Anyway it is still necessary to explore how anisotropy and asymmetry could locally modify the spatial energy density in the Universe. To this aim we have also to take in consideration the whole fundamental asymmetry which, as we ascertained, underlies all interactions at terrestrial level as well at cosmological level. We recall that the violation of LLI can be looked at from three different points of view: kinematical, energetical and from the point of view of spatial asymmetry. In the previous chapter, it has been shown the correlation between the spatial direction of the Cold Spot with respect to the Earth and the direction along which, in terrestrial experiments both electromagnetic and nuclear, it is possible to measures the violation of local Lorentz invariance. In other words, we compared cosmological evidences and terrestrial experiments from the point of view of the spatial asymmetry and anisotropy that appear in both the two apparently unrelated scopes of investigation. In this sense, we have considered the more general concept of Lorentz violation instead of the breakdown of local Lorentz invariance, which is used according to the standard literature, in order to extend this comparison from the geometrical level to the energetic one. We recall that the background cosmology of the universe depends on an expansion parameter, the Hubble constant, and on other several parameters that determine the matter and energy content of the universe. These parameters are expressed in term of the universe critical density $\rho_{\text{crit}} = 1.87847 \cdot 10^{-29}\,\text{g/cm}^3$, defined from the Friedman equation as the density for which the universe is spatially flat, and they govern the physics of the anisotropies of the CMBR and hence the measurement of these anisotropies constraints the values that these parameters may take. In this sense, the value of the critical density expressed as $1.054 \cdot 10^4\,\text{eV/cm}^3$ depends on the CMBR anisotropies and can be compared to the energy densities typical of the the deformed Space-Time phenomena. In fact, as the critical density ρ_{crit} refers to a flat universe, the same happens for the DST densities which depend on an energy threshold which separates the Minkowskian flat space-time from the deformed one. We report here only the energy density values of the gravitational and electromagnetic interactions, since *a priori* one may consider only these two interactions as part of the phenomenon of the CMBR and *a posteriori* since only the energy density of these two interactions is compatible with that of the CMBR if one considers the size range of the Ridolfi's cavities giving the spatial critical volume. Rigorously, the Ridolfi's cavities have been identified in materials where DST nuclear reactions occur. However, due to the novelty of the comparison between the universe energy density and the energy densities to generate a deformed space time for the gravitational and electromagnetic interaction, the volume of these cavities has been used also for these two interactions as a heuristic hypothesis, as already done in Chapter 12.

Following an apocryphal aphorism ascribed to E. Fermi 'In physics a bad number is better than no number'.

Of course this does not exclude the possibility of an energetic compatibility between the hadronic/leptonic interaction and the CMBR. Therefore, we now report in Table 23.1 the values of the critical energy densities for the radius of the Ridolfi's cavities from 4 to 10 microns, again as done in Chapter 12.

For the electromagnetic interaction, the value of the critical energy density for a radius between 4 and 5 microns is compatible with the critical energy density of the universe. A similar compatibility exists with the energy density of the gravitational interaction for a cavity radius between 7 and 8 microns.

This is the same result as in Chapter 12, giving us one more argument of comparison between terrestrial and cosmic phenomena.

The correct comparison appears to be between the critical energy densities of the interactions and the critical energy density of the universe that has a spatially flat geometry according to Friedman. This compatibility of the critical density of the universe both with

Table 23.1. Critical energy densities.

Ridolfi cavity radius (μm)	Spherical volume V (cm^3)	ρ_{grav} gravitational ($E_{0,\text{grav}} = 20.2\,\mu$eV) $\rho_{\text{grav}} = E_{0,\text{grav}}/V$ eV/cm^3	ρ_{el} electromagnetic ($E_{0,\text{grav}} = 4.5\,\mu$eV) $\rho_{\text{el}} = E_{0,\text{el.}}/V$ eV/cm^3
4	$2.68 \cdot 10^{-10}$	$7.53 \cdot 10^{4}$	$1.68 \cdot 10^{4}$
5	$5.24 \cdot 10^{-10}$	$3.86 \cdot 10^{4}$	$8.59 \cdot 10^{3}$
6	$9.05 \cdot 10^{-10}$	$2.23 \cdot 10^{4}$	$4.97 \cdot 10^{3}$
7	$1.44 \cdot 10^{-9}$	$1.41 \cdot 10^{4}$	$3.13 \cdot 10^{3}$
8	$2.14 \cdot 10^{-9}$	$9.42 \cdot 10^{3}$	$2.10 \cdot 10^{3}$
9	$3.05 \cdot 10^{-9}$	$6.62 \cdot 10^{3}$	$1.47 \cdot 10^{3}$
10	$4.19 \cdot 10^{-9}$	$4.82 \cdot 10^{3}$	$1.07 \cdot 10^{3}$

the electromagnetic and gravitational energy density appears to be rather interesting, intriguing and sound, in the framework of DST Theory, where we can consider ρ_{crit} as a boundary between the flat and non flat parts of the Universe. As a matter of fact, the critical density of the universe ρ_{crit} is governed both by the anisotropy values of the CMBR, which is an electromagnetic phenomenon, and by the gravitational interaction which slows down the expansion of the universe contributing to determine the temperature and the anisotropy of the CMBR.

It appears to exist a fairly strong correlation between terrestrial scale experiments and the properties of cosmological scale phenomena:

— Asymmetry in LLI violation
— Anisotropy in DST neutron emission
— Anisotropy in CMBR
— Critical energy density in universe.

23.4 Asymmetry

As a matter of fact, we face up ourselves with a situation we could define analogous to that of the references systems in the four dimensional Minkowskian space-time, where the reference frames, as well known, are Cartesian but non orthogonal, and the angle between the axes, in two dimensions, is given by a trigonometrical relation having as argument the ratio between the velocity of light and the velocity of the moving reference frame itself. This can be a new starting point to consider how the fundamental angle of asymmetry, coming from angle law, can be used for evaluating the energy corresponding to the angle itself. We can consider such an angle as that of a non orthogonal Cartesian frame of a Minkowskian type and then to use it for evaluating to what velocity such angle can be related to and therefore what energy it corresponds to, within a physical system endowed with inertia moving with such velocity. If from one side we can imagine this new way of using the angle law to connect it with an equivalent energy, given a reference physical system, from another side we cannot disregard a fact, physical as well as mathematical at the same time, which we can summarize in the problem of the scalar product. Just now we can define the scalar product in a 'space' understood in its double nature mathematical and physical. A certain 'mathematical space' is endowed with its own scalar product, whereas the 'physical space' is endowed with physical extended coordinates connected by measure relations of 'physical intervals', also called 'distances', given by the scalar product of the mathematical space put in correlation with the physical space itself. Usually the two spaces, the mathematical one and the physical one, are considered as coincident, or 'indistinct', so that we have the following situation. The physical space of Euclides has space coordinates and the mathematical Euclidean space has a scalar product defined positive, which is the Pythagoras theorem, by which the orthogonal projections are measured and above all the spatial distances are measured.

The physical space of Minkowski has space and time coordinates and the mathematical Minkowskian space has a scalar product half-defined positive, by which the projections are measured, non-necessarily orthogonal, due to Cartesian frames non-orthogonal, already quoted above, and by which the space-time distances are measured connected also with the causality relation of the phenomena. This last fact is due to the homogeneity between space and time which demands the existence of a physical entity giving rise just to this homogeneity. Traditionally this entity is a velocity, preferably invariant for reference frame, which assumes the role of maximal causal velocity since connects the cause and the effect of an interaction with a corresponding time interval between the spatial positions of the cause and of the effect. At last the deformed space time, also it with space and time coordinates, which in its corresponding mathematical space has still a scalar product half-defined positive but with an operational definition of the product different from the Minkowskian one. The final consequence to which we have to pay attention can be recapitulated in the question of what is the mathematical way by which the asymmetry affects the scalar product suitable when we want to take care simultaneously of the energy and asymmetry. Now, we look at the problem of how to implement the asymmetry and its angle law within the energy metrics which give the geometrical description of the fundamental interactions. We put forward the possibility to use Heaviside step function to introduce the asymmetry angles in the mathematical representation of the energy metrics. Borrowing the use of step functions from Chapters 1 and 2 we can image to put a step function depending from a suitable angle as variable in each metric element. Obviously this proposal is not so easy to use in both senses mathematical and physical. Firstly, we have to explore if to make this step function depending also from energy thresholds. This can mean that we choose to apply the asymmetry only in the non-flat, deformed, part of the metrics. Otherwise we can assume, from the experimental evidences of the angle law, that the asymmetry works ever in all the dimensions, energy space and time, and therefore we have to put it in all metric element, regardless of any energy range. In this last case, a possible way can be to multiply the metric element by a suitable step function depending from angle. We do not wish to push further this problem since the purpose of this book is to present only the Lorentz violation as due to a fundamental asymmetry and this fact complete the Deformed Space Time Theory in the framework of a five-dimensional picture of natural law based on energy, space and time.

23.5 Methodological Remarks

The very foundation of the Galilean experimental method is strongly connected with the experiment repetition. But this apparently simple concept is evolved during the centuries after Galilei and with the technical and technological facilities at disposal for physics experiment. In this book we presented and exploited the concept of 'repeatability in great' as introduced at the beginning of Chapter 15 and recalled at the beginning of Chapter 19. Now we wish to specify in which methodological context this concept has its place. The three main concepts we deal with within the experimental method are the following: repeatability, reproducibility and repeativity.

Repeatability means that an experiment can be repeated at will whenever and wherever, or rather regardless the place and the time on the Earth surface, giving rise to the same results, but all considered, say the devices used and the laboratory environment. In this way in physics it is possible to assess at least a physical law or at most a natural law. Since natural law can be considered any fundamental interaction ruling what in consequence can be called a 'class of phenomena'. Whereas a physical law can be regarded as a specific consequence of a natural law.

Reproducibility means that an experiment has a commonly known way, layout, devices, materials and methods together with techniques and technologies such to allow to reproduce an experiment in any laboratory by any experimentalist, obviously all this regardless of any patent problem or intellectual property due to original experimentalists firstly performing the experiment. This is the basis on which is grounded the human confidence to relay on any experimental results. It is quite obvious that behind the human soundness about an experiment and its results there are many other issues but they are outside the purposes of this book. Anyway, this ability can give confidence on the actual existence of a physical law or a natural law.

Repeativity means the possibility to repeat a fixed experiment without changing any detail many times at will in order to get data allowing to treat the results by the mathematical methods of statistics.

Each of these concepts has a possible contra-indication, say to pay attention only to a single result could hide the need that a natural phenomenon is better studied in a wide range of results, despite the financial cost that this need represents. Again this could be an issue connected with the human character of the experimentalist and therefore it concerns more the psychology than the physics.

In this book we followed all these prescriptions at different levels but we wanted to do a step more introducing the 'repeatability in great', which means to repeat the occurrence of a law controlling in the same way different phenomena in different physical systems, even ruled by the same interaction, giving rise to the interpretation that this law is fundamental part of the interaction itself, or an independent 'natural law' when it controls in the same way different phenomena ruled by different interactions. We strongly and deeply believe that this concept within the experimental method is a powerful tool in exploring natural situation opposing the nature logic to the human logic, which unfortunately are not the same, the logic of the nature is not the human logic:

ordo et connectio rerum naturae idem non est quam ordo et connectio hominis idearum.

Index

www.ingramcontent.com/pod-product-compliance
Lightning Source LLC
LaVergne TN
LVHW011705200926
848484LV00004B/1074